Rural Geography

Hitesh Saxena

SRISHTI BOOK DISTRIBUTORS
NEW DELHI (INDIA)

ISBN : 978-81-89948-88-7

Published by:
SRISHTI BOOK DISTRIBUTORS
4378/4-B-202, 2nd Floor,
J.M.D. House, Ansari Road,
Daryaganj, New Delhi-110002
E-mail : srishtibookdistributors@yahoo.co.in
Ph. : 011-23284904

***Edition* : 2012**

Rs. 1100/-

Typeset by :
Shubh-Deep Laser Printer
Delhi-110002

Printed at
Shivani Art Press
Delhi-110032

PREFACE

Rural geography is a long-established area of study within the discipline of Geography. It covers aspects of society, economy, environment and culture in the countryside. In recent years, geographers have given particular attention to change in the countryside which is recognised as being influenced by broader processes that are at work at international and global levels.

Human geography broadly differs from physical geography in that it has a greater focus on studying intangible or abstract patterns surrounding human activity and is more receptive to qualitative research methodologies. It encompasses human, political, cultural, social and economic aspects of the social sciences.

Rural development may be defined as the development of regions excluding the urban areas such as the towns and cities. Smaller settlements such as villages, farmsteads, and market towns are normally included within the concept of rural, while most of the land area is expected to be used as agricultural land, forest or in its natural state. The concept of rural development has changed significantly during the last three decades.

While compiling research material for this book, we surf many concerned websites, blogs for ensuring the authenticity of the matter and for checking facts and figures. All the resources have been acknowledged in bibliography section of this book. We are sincerely indebted to all these resources and librarians who provided books of various authors on this subject topic.

—Hitesh Saxena

CONTENTS

1

Introducing Rural Geography

INTRODUCTION

Trying to define what "rural" is, in a nation of such diverse geography, can be a daunting task; and one's concept of the term can be a bit ambiguous and vague. Particularly, in this age of instant communication, when a Montana rancher can access the New York stock market quotes from his laptop, miles from the nearest town, defining rural becomes more of a subjective term than what is defined by Webster's.

However, for those concerned with rural health care and human services, that which constitutes rural must not be subjective, but rather precise in fulfilling the definition. Federal and state policy makers, as well as service providers and researchers, need a clearly stated definition that is current in its interpretation.

There are three government agencies whose definitions of what is rural are in wide use: the U.S. Census Bureau, the Office of Management and Budget, and the Economic Research Service of the U.S. Department of Agriculture (USDA). They and other organizations continue to strive for more precise definitions to fit new programs as the demographics of the United States are constantly changing. The number of rural counties fluctuates over time, and disparities with old designations continually exist.

The need for a clearer definition to meet the needs of new programs and new policies has encouraged other agencies to create more detailed definitions such as found in the collaboration between the WWAMI Rural Health Research Center and the Economic Research Service of the USDA. Agencies involved with rural health and human services will continue to evolve and adapt themselves, striving to better serve the needs of the rural population, for what is rural today will most likely change as we move on into the new millennium.

Rural Geography is defined as the study of the rural landscapes of the developed world. It includes the origin, development, and distribution of rural settlement, rural depopulation, the causes and consequences of agricultural change, patterns of recreational use of the countryside, tourism, planning, and the growing influence on rural areas of urban dwellers. It is a kind of Human Geography.

MEASURING RURALITY: WHAT IS RURAL?

According to official U.S. Census Bureau definitions, rural areas comprise open country and settlements with fewer than 2,500 residents. Urban areas comprise larger places and densely settled areas around them. Urban areas do not necessarily follow municipal boundaries. They are essentially densely settled territory as it might appear from the air. Most counties, whether metropolitan or nonmetropolitan, contain a combination of urban and rural populations.

Urban areas are of two types—urbanized areas and urban clusters—identical in the criteria used to delineate them but different in size. The Census Bureau defines an urbanized area wherever it finds an urban nucleus of 50,000 or more people. They may or may not contain any individual cities of 50,000 or more (152 currently do not). In general, they must have a core with a population density of 1,000 persons per square mile and may contain adjoining territory with at least 500 persons per square mile. Urbanized areas have been delineated using the same basic threshold (50,000 population) for each decennial census since 1950, but procedures for delineating the urban fringe are more liberal today. In 2000, 68 percent of Americans lived in 452 urbanized areas.

The same computerized procedures and population density criteria are used to identify urban clusters of at least 2,500 but less than 50,000 persons. This delineation of built-up territory around small towns and cities is new for the 2000 census. In 2000, 11 percent of the U.S. population lived in 3,158 urban clusters.

According to this system, rural areas consist of all territory located outside of urbanized areas and urban clusters. The U.S. rural population was 59 million (21 percent) in 2000.

ERS researchers and others who discuss conditions in "rural" America most often refer to conditions in nonmetropolitan areas. Metropolitan (metro) and nonmetropolitan (non-metro) areas are defined on the basis of counties. Counties are typically active political jurisdictions, usually have programmatic importance at the Federal and State level, and estimates of population, employment, and income are available for them annually. They are also frequently used as basic building blocks for areas of economic and social integration.

Metro and non-metro areas are defined by the Office of Management and Budget (OMB). In 2003, OMB defined metro areas as (1) central counties with one or more urbanized areas, and (2) outlying counties that are economically tied to the core counties as measured by work commuting. Outlying counties are included if 25 percent of workers living in the county commute to the central counties, or if 25 percent of the employment in the county consists of workers coming out from the central counties—the so-called "reverse" commuting pattern. Non-metro counties are outside the boundaries of metro areas and are further subdivided into two types: micropolitan areas, centered on urban clusters of 10,000 or more persons, and all remaining "noncore" counties.

Federal data for certain social and economic characteristics of counties are available on an annual basis, some even more frequently. In contrast, data on the characteristics of rural and urban residents are available only from the decennial censuses. Using population counts from the 2000 Census, the following table shows the number of residents of rural and urban areas versus nonmetro and metro areas. There were 59.1 million rural residents in 2000, a little less than half (49 percent) of whom lived in nonmetro counties. There were 49.2 million nonmetro county residents, 59 percent of whom lived in rural areas. Metro county residents are preponderantly urban area residents—87.1 percent. Overall,

17 percent of the national population lived in nonmetro counties and 21 percent lived in rural areas in 2000. For the first time, a slight majority of rural people now live in metro areas.

NATURE AND SCOPE OF RURAL GEOGRAPHY

Human geography broadly differs from physical geography in that it has a greater focus on studying intangible or abstract patterns surrounding human activity and is more receptive to qualitative research methodologies. It encompasses human, political, cultural, social and economic aspects of the social sciences. While the major focus of human geography is not the physical landscape of the Earth, it is not possible to discuss human geography without going into the physical landscape, on which human activities are being played out and environmental geography is emerging, as an important link between the two. Human geography is methodologically diverse, using both qualitative methods and quantitative methods, including case studies, survey research, statistical analysis and model building, among others. Thematically, human geography may be concerned with an array of human enterprises, from villages and cities, schools, health, commerce and trade, to name a few. The spatial human architecture of a variety of institutions and practices unites these entities within the discipline. For example, a human geographer might be concerned with the geographic patterns of communicable diseases, school performance in rural versus urban school districts or the rise of innovative technology clusters.

Bottom of Form

Rural Geography is a long-established area of study within the discipline of Geography. It covers aspects of society, economy, environment and culture in the countryside. In recent years, geographers have given particular attention to change in the countryside which is recognised as being influenced by broader processes that are at work at international and global levels. Various concepts and models have been developed in order to gain better understanding of these processes. A key objective of this course is to promote better awareness of and insight into change in the European countryside. The course is designed around seven key themes relating to rural society and economy. Three lectures deal with each theme and each set of lectures begins with discussion of concepts relating to the theme. Practical examples are provided from a range of European countries, including Ireland.

The following are some of the major features of change in the appearance of countryside and in the occupations and lifestyles of rural dwellers. The boundaries between the 'rural' and the 'urban' have become blurred, for example, through the encroachment of residential settlement, industrial spaces and recreational facilities on rural land in the hinterland of towns and cities. Certain parts of the countryside have a long history of out-migration associated with the limited employment opportunities available. Other areas, by contrast, have experienced increased immigration in recent decades of both national and international migrants on the margins of cities and towns and even in more remote locations. Rural residents are being assigned new responsibilities for meeting their own social and economic needs in many areas, as the central state becomes increasingly unable to support employment and services. Attention is often drawn to the decline of farming and fishing, and the need to adopt alternative activities to supplement incomes. The promotion of manufacturing industry is one such long-established alternative. New information and communication technologies offer other opportunities. Rural land use is changing also and new products and processes are being introduced in the wake of continuing reform of the CAP. At the same time, the countryside remains distinctive in the minds of many rural and urban dwellers and is marketed as a central component of national tourism in many countries.

COMPARE RELATIONSHIP BETWEEN URBAN GEOGRAPHY AND RURAL GEOGRAPHY

Urban areas have a large number of buildings and people in a small area, and rural areas have few buildings and people in a large area. Rural areas have a population of less than 5,000, and urban areas have a population of more than 5,000. Human settlements are classified as rural or urban depending on the density of human-created structures and resident people in a particular area. Urban areas can include town and cities while rural areas include villages and hamlets.

While rural areas may develop randomly on the basis of natural vegetation and fauna available in a region, urban settlements are proper, planned settlements built up according to a process called urbanization. Many times, rural areas are focused upon by governments and development agencies and turned into urban areas.

Unlike rural areas, urban settlements are defined by their advanced civic amenities, opportunities for education, facilities for transport, business and social interaction and overall better standard of living. Socio-cultural statistics are usually based on an urban population. While rural settlements are based more on natural resources and events, the urban population receives the benefits of man's advancements in the areas of science and technology and is not nature-dependent for its day to day functions. Businesses stay open late into the evenings in urban areas while, sunset in rural areas means the day is virtually over.

The flip side of this is that rural areas do not have pollution or traffic problems that beset regular urban areas. Many governments, though focusing on the development of rural areas, have also tried to 'protect' these areas as preservation of their country's basic culture and traditions. Urban areas are also classified according to land use and density of population. But this can vary from developed countries to developing countries. For example, in Australia, urban cities must include at least a 1,000 residents with 200 or more people per square kilometer while in Canada, an urban area is defined with a density of 400 people per square kilometer In China, the density requirement for an urban area is about 1,500 people per square kilometer Statistically, two urban areas with less than two kilometers between them are considered one urban zone.

THE TERM "HEMLET"

A hamlet is a type of settlement. The definition of hamlet varies by country. It usually refers to a small settlement in a rural area, or a component of a larger settlement or municipality. Hamlets are typically unincorporated communities.

Multifunctional Rural Geographies : Reactionary or Radical?

Recent developments, empirical and theoretical, call into question the geographic categories and concentrations of rural geography. This review is written in a world in which, for the first time ever, the urban population exceeds the rural. Davis (2004) emphasizes that this watershed marks not a mere proportional shift in population between static categories, but the continuation of global dynamics in which the rural and urban are mutually constituted by processes that span both. Murdoch and Lowe (2003) make a similar point from a Latourian perspective, showing that attempts to draw and maintain sharp boundaries between the rural and urban actually accelerate processes that transgress those boundaries and link the two.

Yet, while such processes are at work around the world, 'rural' geography as such continues to focus overwhelmingly on the rural areas of industrialized Anglophone countries, leaving investigation of most of the rural world to scholars of development, postcolonialism, natural resource industries, and other areas of the discipline (Adriansen and Madsen, 2004). The few efforts to apply theory from 'rural geography' directly to rural areas and processes in the global South (e.g., Wilson and Rigg, 2003) are all the more notable precisely because they are so rare. 'Rurality' thus remains at least as much a product of divisions of labor within the academy and social contexts as a category defined by particular sectoral mixes, land uses, densities, or other empirical descriptors. Progress continues on several topics that have become central to rural geography in recent years. Thus, new work examines insights produced by the cultural turn in rural geography (e.g., Holloway and Kneafsey, 2004a; Phillips, 2004), exclusions in rural societies (e.g., Parr et al., 2004; Woods, 2004), alternative food networks (e.g., Holloway and Kneafsey, 2004b), and issues of governance in rural areas, particularly with respect to the role of communities (e.g., Edwards and Woods, 2004; Seymour, 2004).

The empirical and analytical validity of 'postproductivism' continues to be a major topic of debate, as discussed later in this review. Recent overviews and examples of work on all of these themes can be found in Beesley et al. (2003) and Holloway and Kneafsey (2004a). This review will focus on a single important theme in recent rural geography, that of 'multifunctionality' in rural landscapes. Multifunctionality, the idea that rural landscapes typically produce a range of commodity and noncommodity use values simultaneously and that policy ought to recognize and protect that entire range of values, has received substantial attention in the field in recent years. It figures in the debates concerning governance and postproductivism referenced above, for example. It has also been widely debated in cognate fields, including rural sociology, agricultural economics, and environmental economics. Recent work has examined multifunctionality's robustness, coherence, and geographic applicability, while also proposing competing theoretical framings of it. The review also suggests potential connections to current work on the neoliberalization of environmental governance and environmental indicators (McCarthy and Prudham, 2004; Liverman, 2004). Situating research on multifunctionality in this way begins to bridge the surprising gap, identified by Milbourne (2003), between the literatures on social nature and on rural geographies. It also responds to Little' (2001: 98) call in these pages to link 'specific policy areas [to] broader trends and processes in the relationship between the economy, the state, policy, and society.' Finally, areas for future research are identified.

From Postproductivism to Multifunctionality

Multifunctionality seems poised to succeed postproductivism as a framework within which to interrogate contemporary rural dynamics. Productivist rural landscapes supposedly centered on high-intensity production of a relatively small range of primary commodities. Increasingly, though, demands on rural areas extend beyond such production and include demands for the provision of ecosystem services, amenities and aesthetics, and preservation of cultural landscapes. Many rural geographers have characterized these trends as a shift towards postproductivism, a framing that has been criticized on both empirical and theoretical grounds. Critics have pointed out, for instance, that rural commodity production remains substantial, that the thesis fits countries outside western Europe poorly if at all, and that the areal differentiation at multiple scales presumably central to any geographic inquiry would seem to be glossed over in a theory framed in terms of an epochal shift (Wilson, 2001; Evans et al., 2002; Goodman, 2004; Wilson, 2004). In response, a number of authors have begun to interpret the trends identified in work on postproductivism as a shift towards multifunctionality in rural areas instead (e.g.,Wilson, 2001; Lowe et al., 2002; Holmes, 2002; Wilson and Rigg, 2003; Wilson, 2004).

As Potter and Burney (2002: 35) summarize it, 'agriculture is multifunctional, producing not only food but also sustaining rural landscapes, protecting biodiversity, generating employment and contributing to the viability of rural areas'. This framing has several advantages over 'postproductivism: it offers a positive characterization rather than a negative one; it recognizes the continued importance of commodity production in rural areas; and it is inherently sensitive to spatial and social differentiation, thefact that different rural areas clearly can and will produce very different, even unique, combinations of use values. A genealogy of 'multifunctionality' makes clear that it is a product ofneoliberal reforms. The term emerged out of the 1994 Uruguay Round Agreement on Agricultural Trade, which integrated agriculture into the ongoing liberalization of international trade (Losch, 2004). Member countries committed to reducing agricultural subsidies, but those deemed 'non-trade-distorting', or not linked to commodity production, were exempt, protected in a new 'green box' safe from liberalization (Potter and Burney, 2002; Hollander, 2004; Dobbs and Pretty, 2004).

The EU and a few allied countries first advanced the notion of multifunctionality in this context, arguing that their payments to rural producers were not direct subsidies of commodity production, but rather payments for the range of noncommodity goods and services jointly produced in the process of commodity production, or support for broader national goals such as food security and rural development. The concept thus emphasizes the positive externalities produced by agriculture rather than the negative ones (Hollander, 2004), and many would say it applies to other natural resource industries as well. Some have argued that these positive externalities are disproportionately produced by more marginal producers -for example, that farms that are less intensive and more diverse are more likely to provide wildlife habitat and valued cultural landscapes. Since such producers are precisely the ones most threatened by trade liberalization, policies to protect multifunctionality often disproportionately enroll economically marginal producers. This has been done in part by making payments more area-based, for example (Bills and Gross, 2005). Still, payments are not entirely decoupled from commodity production, the usual standard for inclusion in the 'green box' (Potter and Burney, 2002).

Multifunctionality has been elaborated in the context of policy struggles over ongoing liberalization. It has become a major issue in reforms of the EUs Common Agricultural Policy, in OECD policy, and in recent WTO negotiations, with coalitions of countries forming to support or oppose it. It is thus not surprising that debates about multifunctionality and its uses have proliferated rapidly in rural geography.

URBAN VS RURAL INDIA

About 80 per cent of the Indian population live in villages. When travelling through the length and breadth of this subcontinent, one can really visualise the difference between rural and urban India.

There is a big difference between urban and rural India. One of the major differences that can be seen between rural India and urban India, is their standards of living.

People living in urban India have better living conditions than those living in the rural parts of India. There is a wide economic gap between rural and urban India. Rural India is very poor when compared to Urban India.

Another difference that can be seen between urban and rural India, is their education. In rural India, the parents seldom educate their children, and instead, make their children work in the fields. Poverty, and lack of sufficient infrastructure, can be attributed to the lack of education in rural India.

When considering homes, about three-quarters of the households in urban India live in pucca homes. On the other hand, only a quarter of the people in rural India live in pucca homes. While groundwater is the main source of drinking water in rural India, the urban people rely more on tap-water.

Urban India is almost electrified when compared to rural India. One can even come across villages where electric power is not yet available. When comparing the sanitary facilities, it is limited in rural india. About 90 per cent of the households in rural India do not have latrines, but this is not the case in urban India.

Most of the developments have not yet reached the rural parts of India. With regards to health care as well, rural India lacks good hospitals when compared to urban India. Some of the rural areas even lack a dispensery.

1. People living in urban India have better living conditions than those living in the rural parts of India.
2. Rural India is very poor when compared to urban India.

3. In rural India, the parents seldom educate their children, and instead, make their children work in the fields.
4. About three-quarters of the households in urban India live in pucca homes. On the other hand, only a quarter of the people in rural India live in pucca homes.
5. Most of the developments have not yet reached the rural parts of India.
6. With regards to health care, rural India lacks good hospitals when compared to urban India.

HISTORY OF URBAN AND RURAL AREAS

Statistics have been split into urban and rural categories in decennial census publications for over a century. The definition of "urban" has changed over time in response to changes in settlement patterns, data use needs, and technology available for use in defining urban areas. The Census Bureau has continued to define "rural" as all territory, persons, and housing units not defined as urban. In the censuses of 1880, 1890, and 1900, places were deemed urban based on minimum population sizes of 8,000, 4,000, and 2,500 inhabitants.

Beginning in 1910, the minimum population threshold to be categorized as an urban place was set at 2,500. "Urban" was defined as including all territory, persons, and housing units within an incorporated area that met the population threshold. The 1920 census marked the first time in which over 50 percent of the U.S. population was defined as urban.

The Census Bureau revised the urban definition for the 1950 census by adopting the urbanized area concept, to better account for increased growth in suburban areas outside incorporated places of 50,000 or more population. This change made it possible to define densely-populated but unincorporated territory as urban. The Census Bureau continued to identify as urban those places that had populations of 2,500 or more and were located outside urbanized areas. The Census Bureau also officially identified unincorporated places (referred to as census designated places (CDPs) starting with the 1980 census) located outside urbanized areas for the first time in 1950, and designated as urban any that contained at least 2,500 people within its boundaries. In 1960, the Census Bureau also adopted a population density threshold of at least 1,000 people per square mile for urbanized areas.

For Census 2000, the Census Bureau adopted the urban cluster concept, for the first time defining relatively small, densely settled clusters of population using the same approach as was used to define larger urbanized areas of 50,000 or more population, and no longer identified urban places located outside urbanized areas. In addition, all urbanized areas and urban clusters were delineated solely on population density, without reference to place boundaries (for the 1950 through 1990 censuses, places were included in, or excluded from, urbanized areas in their entirety; exceptions were made for incorporated places containing substantial amounts of sparsely populated territory).

GLOBALIZATION IMPACT ON EMPLOYMENT, EDUCATION AND RURAL CULTURE IN INDIA

Globalization has affected most the earnings and the living standards of people in the urban areas. NGO's and several organizations are studying the impact of globalization on value system and employment. People always discuss its impact in relation to literacy, skills, growth and development mostly on urban life but its effect on rural livelihood where more than 60 percent of Indian population resides is seldom discussed.

Rural infrastructure: In Indian villages, people still live in houses made of mud barring houses of few rich and progressive farmers. Government made houses for people in the name of "Indira Awaas Yojna" are poorly designed and constructed. These houses are in no way better than the houses in which poor village people are already residing. Dwelling units made in the name of "Indira Awaas" are so small that a family of three to four people can not live comfortably in these houses. This appears sheer wastage of people's money in the name of subsidy for helping poor people. The worst aspect of entire episode is location of these houses. People in villages prefer to stay in clusters and large spaced houses. They can stay in muddy houses but they can not stay in clumsy and one or two room houses. They may reside in open space but not in space constrained houses. Occupation has been taken by the people but most of them are still staying in their earlier made "Kuchcha " houses. Farming communities are yet untouched from the wind of globalization.

Labor class has migrated to cities in search of employment. Scarcity of labor has further worsened by the 'minimum wage act'.

Earlier, wages were determined by different factors like demand and supply, nature of work and efficiency of the laborers.Poor farmers finding it difficult to hire workers on the wages determined by government and its ministries. Fund paucity has compelled them to perform farming activities using family labor. Government never allows petrol and crude prices to be determined by the market forces. Likewise, framing a policy of minimum wages to daily paid labors is an easy way to show Govt.'s sympathy to poor people.

Government has initiated several developmental programs for uplifting of living standards of poor people but full benefits have not reached to the targeted population due to corruption prevalent in administrative and political systems.Pradhan Mantri Gramin Sadak Yojna has resulted in road connectivity in rural India but roads are of poor quality and without drainage support. Toilet and lavatory systems are not of standard quality and not even constructed in all houses of the village. Even today more than 90 percent people in villages attend the call of nature in open fields. Electricity: Life in rural India is miserable due to non-availability of electricity. Several states in India claim that 40, 50 or even 100 percent villages have been electrified. But supply of electricity to villages that have been electrified is not more than 3-4 hours per day. It is big hindrance in development. Globalization is not going to make much difference to rural life until ad unless electricity is supplied uninterruptedly 10-12 hours per days too these villages.

Education: School buildings are available in few villages but number of teachers is inadequate in primary schools. Benches, boards and other facilities are of sub-standard quality. There is, however, one positive development that girls are attending the schools in the villages. Also the number of students attending graduate and post graduate courses is increasing but awareness among students from rural areas lacks towards technical education and that is the single reason that most of the students from rural areas are unable to secure employment.

Technology: Technology has failed to percolate to villages in absence of electricity and other communication infrastructure. Few people know about the internet. However, well-to-do families have availed DTH and dish TV facilities. Mobile connections are increasing in rural areas but at slower pace. There are no small scale industries in villages to provide employment to educated youth.

Culture and social values: India's real culture is still preserved in rural life. New advancement of technology has not much influence in rural areas. People still prefer to wear dresses of old fashion and celebrate festivals in old styles. Folk dances and folk songs are still popular among villagers. Culture is still untouched and unaffected by western influence. Globalization has no impact on rural life as standards of living are suboptimal but migration of people is taking place and poor people are moving to urban areas in search of employment.

Agriculture: Globalization does not have any positive impact on agriculture. On the contrary, it has few determental effect as government is always willing to import food grains, sugar etc whenever there is a price increase of these commodities. Government never thinks to pay more to farmers so that they produce more food grains but resorts to imports. On the other hand, subsidies are declining so cost of production is increasing. Even forms producing fertilizers have to suffer due to imports. There are also threats like introduction of GM crops, herbicide resistant crops etc.

CHARACTERISTICS OF THE RURAL LANDSCAPE

A classification system of eleven characteristics has been developed for reading a rural landscape and for understanding the natural and cultural forces that have shaped it. Landscape characteristics are the tangible evidence of the activities and habits of the people who occupied, developed, used, and shaped the land to serve human needs; they may reflect the beliefs, attitudes, traditions, and values of these people.

The first four characteristics are processes that have been instrumental in shaping the land, such as the response of farmers to fertile soils. The remaining seven are physical components that are evident on the land, such as barns or orchards. Many, but not all, rural properties contain all eleven characteristics. When historic processes are linked to existing components, the rural landscape can be viewed as a unified whole. The section "Documentation of Landscape Characteristics" shows the relationship of the eleven characteristics and the features represented by them.

This classification system is a tool for gathering and organizing information. First of all, it is used to develop historic contexts for rural areas. The processes define specific themes, such as dairy farming or Belgian settlement, that have influenced historic development. The physical components define historic features of the landscape that may be used to describe significant property types and to identify properties eligible for listing in the National Register.

Second, the system is used to identify and evaluate the significant properties of a rural area or to determine the eligibility of a particular rural landscape. Through field survey and historic research, characteristics are associated with specific features, such as field patterns or roadways, and provide an understanding of an area or property's historic land uses and physical evolution.

Third, as information about existing characteristics is related to the historic contexts for a geographical area, assessments of significance, integrity, and boundaries can be made for specific properties. Information is evaluated to determine whether, within a rural area or region, a large historic district or separate properties should be considered for listing in the National Register. A comparison of past and present characteristics within a single property helps determine whether the property retains historic integrity and what the National Register boundaries should be.

Finally, the classification system provides a format for documenting rural properties on National Register forms. It can be used to organize the description and statement of significance for a specific rural property on the registration form. It is also useful for organizing information about rural historic contexts and property types on the multiple property documentation form.

Processes

1. Land Uses and Activities: Land uses are the major human forces that shape and organize rural communities. Human activities, such as farming, mining, ranching, recreation, social events, commerce, or industry, have left an imprint on the landscape. An examination of changing and continuing land uses may lead to a general understanding of how people have interacted with their environment and provide clues about the kinds of physical features and historic properties that should be present.

Topographic variations, availability of transportation, the abundance or scarcity of natural resources (especially water), cultural traditions, and economic factors influenced the ways people use the land. Changing land uses may have resulted from improved technology, exhausted soils or mineral deposits, climatic changes, and new economic conditions, as well as previous successes or failures. Activities visible today may reflect traditional practices or be innovative, yet compatible, adaptations of historic ones.

2. Patterns of Spatial Organization: The organization of land on a large scale depends on the relationship among major physical components, predominant landforms, and natural features. Politics, economics, and technology, as well as the natural environment, have influenced the organization of communities by determining settlement patterns, proximity to markets, and the availability of transportation.

Organization is reflected in road systems, field patterns, distance between farmsteads, proximity to water sources, and orientation of structures to sun and wind. For example, spatial patterns can be seen in the grid of square mile townships and 160-acre farmsteads in the Midwest established by the land ordinances of 1785 and 1787; the distribution of towns every seven miles along a railroad corridor; and the division of land in Louisiana, by the French long-lot system, to ensure that every parcel has river frontage.

Large-scale patterns characterizing the settlement and early history of a rural area may remain constant, while individual features, such as buildings and vegetation, change over time. Changes in technology, for example, may have altered plowing practices, although the location of plowed fields, and, therefore, the overall historic pattern may remain the same.

3. Response to the Natural Environment: Major natural features, such as mountains, prairies, rivers, lakes, forests, and grasslands, influenced both the location and organization of rural communities. Climate, similarly, influenced the siting of buildings, construction materials, and the location of clusters of buildings and structures. Traditions in land use, construction methods, and social customs commonly evolved as people responded to the physiography and ecological systems of the area where they settled.

Early settlements frequently depended upon available natural resources, such as water for transportation, irrigation, or mechanical power. Mineral or soil deposits, likewise, determined the suitability of a region for particular activities. Available materials, such as stone or wood, commonly influenced the construction of houses, barns, fences, bridges, roads, and community buildings.

4. Cultural Traditions: Cultural traditions affect the ways that land is used, occupied, and shaped. Religious beliefs, social customs, ethnic identity, and trades and skills may be evident today in both physical features and uses of the land. Ethnic customs, predating the origins of a community, were often transmitted by early settlers and perpetuated by successive generations. Others originated during a community's early development and evolution. Cultural groups have interacted with the natural environment, manipulating and perhaps altering it, and sometimes modifying their traditions in response to it.

Cultural traditions determined the structure of communities by influencing the diversity of buildings, location of roads and village centers, and ways the land was worked. Social customs dictated the crops planted or livestock raised. Traditional building forms, methods of construction, stylistic finishes, and functional solutions evolved in the work of local artisans. For example, rustic saunas appeared among the outbuildings of Finnish farmsteads in northwestern Michigan, while community churches occupied isolated crossroads in the High Plains. Taro, grown as a staple in the Hawaiian daily diet, also assumed an important role in the traditional luau. At the Amana Colonies in Iowa, large expanses of farmland and forest—based upon communal ownership, a village settlement pattern, and religious beliefs—varied from the rectangular grid typical of Midwestern family farms.

Components

Circulation Networks: Circulation networks are systems for transporting people, goods, and raw materials from one point to another. They range in scale from livestock trails and footpaths, to roads, canals, major highways, and even airstrips. Some, such as farm or lumbering roads, internally served a rural community, while others, such as railroads and waterways, connected it to the surrounding region.

Boundary Demarcations: Boundary demarcations delineate areas of ownership and land use, such as an entire farmstead or open range. They also separate smaller areas having special functions, such as a fenced field or enclosed corral. Fences, walls, tree lines, hedge rows, drainage or irrigation ditches, roadways, creeks, and rivers commonly marked historic boundaries.

Vegetation Related to Land Use: Various types of vegetation bear a direct relationship to long-established patterns of land use. Vegetation includes not only crops, trees, or shrubs planted for agricultural and ornamental purposes, but also trees that have grown up incidentally along fence lines, beside roads, or in abandoned fields. Vegetation may include indigenous, naturalized, and introduced species.

While many features change over time, vegetation is, perhaps, the most dynamic. It grows and changes with time, whether or not people care for it. Certain functional or ornamental plantings, such as wheat or peonies, may be evident only during selected seasons. Each species has a unique pattern of growth and life span, making the presence of historic specimens questionable or unlikely in many cases. Current vegetation may differ from historic vegetation, suggesting past uses of the land. For example, Eastern red cedars or aspens indicate the natural succession of abandoned farmland in the Midwest.

Buildings, Structures, and Objects: Various types of buildings, structures, and objects serve human needs related to the occupation and use of the land. Their function, materials, date, condition, construction methods, and location reflect the historic activities, customs, tastes, and skills of the people who built and used them.

Buildings—designed to shelter human activity—include residences, schools, churches, outbuildings, barns, stores, community halls, and train depots. Structures—designed for functions other than shelter—include dams, canals, systems of fencing, systems of irrigation, tunnels, mining shafts, grain elevators, silos, bridges, earthworks, ships, and highways. Objects—relatively small but important stationary or movable constructions—include markers and monuments, small boats, machinery, and equipment.

Rural buildings and structures often exhibit patterns of vernacular design that may be common in their region or unique to their community. Residences may suggest family size and relationships, population densities, and economic fluctuations. The repeated use of methods, forms, and materials of construction may indicate successful solutions to building needs or demonstrate the unique skills, workmanship, or talent of a local artisan.

Clusters: Groupings of buildings, fences, and other features, as seen in a farmstead, ranch, or mining complex, result from function, social tradition, climate, or other influences, cultural or natural. The arrangement of clusters may reveal information about historical and continuing activities, as well as the impact of varying technologies and the preferences of particular generations. The repetition of similar clusters throughout a landscape may indicate vernacular patterns of siting, spatial organization, and land use. Also, the location of clusters, such as the market towns that emerged at the crossroads of early highways, may reflect broad patterns of a region's cultural geography.

Archeological Sites: The sites of prehistoric or historic activities or occupation, may be marked by foundations, ruins, changes in vegetation, and surface remains. They may provide valuable information about the ways the land has been used, patterns of social history, or the methods and extent of activities such as shipping, milling, lumbering, or quarrying. The ruins of mills, charcoal kilns, canals, outbuildings, piers, quarries, and mines commonly indicate previous uses of the land. Changes in vegetation may indicate abandoned roadways, homesites, and fields. The spatial distribution of features, surface disturbances, subsurface remains, patterns of soil erosion and deposition, and soil composition may also yield information about the evolution and past uses of the land.

Small-scale elements: Small-scale elements, such as a foot bridge or road sign, add to the historic setting of a rural landscape. These features may be characteristic of a region and occur repeatedly throughout an area, such as limestone fence posts in Kansas or cattle gates in the Buffalo River Valley of Arkansas. While most small-scale elements are long-lasting, some, such as bales of hay, are temporal or seasonal. Collectively, they often form larger components, such as circulation networks or boundary demarcations. Small-scale elements also include minor remnants—such as canal stones, road traces, mill stones, individual fruit trees, abandoned machinery, or fence posts—that mark the location of historic activities, but lack significance or integrity as archeological sites.

✱✱✱

2

Role of Integrated Approach to Rural Development

INTRODUCTION

The objective of this chapter is to outline the role of integrated rural development with special reference to mountain areas. Such an approach is particularly appropriate for mountain regions because of the special problems and opportunities arising from altitude, difficulties of access, micro-climate and the environment of such regions. The nature of both the problems and the opportunities in mountain areas require that a multi-dimensional approach be adopted in order to bring about significant increases in income and improvements in living standards and the quality of life.

Rural development may be defined as the development of regions excluding the urban areas such as the towns and cities. Smaller settlements such as villages, farmsteads, and market towns are normally included within the concept of rural, while most of the land area is expected to be used as agricultural land, forest or in its natural state. What determines the size of settlement to be included within the term "rural" varies from country to country.

Whatever the definition, integrated rural development requires that the influence of the towns not be ignored. The inter-relationship between town and country has to be studied, the problems identified and improvements made where required. Towns are important for providing goods and services, markets for local produce, transportation, storage and processing facilities, social services and recreation for the rural population. The towns and cities within whose hinterland the mountain area falls cannot be ignored.

Though rural development by definition requires that priority be given to rural areas, and the major proportion of investment be made in the villages and countryside, it is important that the linkages with towns be considered and appropriate action be taken to remove obstacles and improve the services provided to rural populations.

INTEGRATED RURAL DEVELOPMENT

The verb "to integrate" is defined by the Oxford dictionary as meaning "complete (imperfect thing) by the addition of the parts", or to "combine (parts) into a whole". It does not involve the collection of a large number of unrelated parts, but of parts which combine together to form a whole. The parts, therefore, should be "linked" and be capable of being "bonded" together within the context of the project. In integrated rural development the range of possibilities for development are examined comprehensively, and the emphasis is on the identification of linkages that relate to the objectives determined. This differs from a strictly agricultural approach, which may ignore the role of other sectors in the development process. For example linkages may be found in respect to the following: .

1. Objectives: For example to stop immigration the welfare services, should be improved.
2. Opportunities: For example the realization of opportunities for tourist development requires improved roads surgeries, cultural centres etc.
3. Cost Reduction: For example abandoned school buildings can be used for health
4. Efficiency: For example improved roads reduce marketing costs and travel time.
5. Needs Satisfaction: For example hotels catering for tourists may also provide recreation for local people if incomes can be improved.

AREA DEVELOPMENT SCHEMES

Integrated rural development schemes are generally "area development schemes", that is they are planned and executed to serve the perceived needs and opportunities of a specific geographical area.

The geographical area of the project or scheme may be determined on the basis of various criteria, such as a region, administrative district, watershed, homogeneous region, physical region (i.e mountain area, plateau etc.), cultural region, ethnic area, agro-climatic region, and economic region (i.e. low income area, depressed region, high unemployment, industrial region etc.). It is generally useful if the project area is easily definable by some commonly accepted criteria or is widely identified as being a region with special characteristics. But even if this is not the case, experience has shown that such projects can be successful even in regions that cannot be identified by clear cut criteria, as long as the demarcation of the area is clearly defined.

Integrated rural development can be defined as "area development schemes" which involve a broad range of activities designed to improve production, infrastructure, services and living standards in line with the objectives, and with emphasis on the linkages between the various components and geographical areas internal and external to the area. This does not mean, however, that everything should be done at the same time, for the determining of priorities and the time sequence of components is important, and the funds available are normally insufficient for everything to be done all at once. Neither does it mean that a project may be described as being "integrated" if a welfare component is simply added to a productive component. For example an irrigation scheme with an education and health component added to it does not constitute an integrated project.

ADVANTAGES OF IRD

Integrated rural development involves a concerted effort to encourage the development of delineated rural areas, in line with national programmes and objectives, but entailing a concentrated effort over and above national programmes. The advantages over national programmes are:

- The concentration of effort for improvements in one area at a time.

- The design of programmes and the combination of schemes specifically suited to conditions in the area, which relate to the problems, potential and conditions of the region.
- The strengthening of local institutions and the encouragement of further (self-sustained) growth.
- The possibilities for cost effectiveness arising from3he concepts of comparative advantage and the identification of linkages between sectors.

DISADVANTAGES OF IRD

The disadvantages are the high cost in terms of national budgets, the budget constraints inhibiting the repetition of the experience in other regions, and the problem of determination of priorities between regions. Because of these factors repeat projects in the same area are difficult because other regions are anxious to get their turn for attention and investment. In general the investment involved in concentrating on one region at a time poses exceptional pressure on the budget because national programmes for rural development cannot be cufiailed, and consequently expenditure over and above is required for the development of specific areas. Another major disadvantage is the complexity of integrated projects which require inputs from many different departments, agencies and organizations. Good co-ordination is therefore the key to success both with respect to preparation and implementation.

STRONG GOVERNMENT COMMITMENT

These budget constraints require that there be strong national commitment for rural development projects to ensure that the necessary funding and other resources needed for successful implementation within the planned timetable are made available. In this, the project approach differs from that of national programmes, because projects have to be implemented within a given time schedule. Any increase in the time needed between investment expenditure and returns to the project, reduces viability and therefore influences the degree of success that can be claimed. Projects funded by international agencies such as the World Bank and the European Investment Bank generally have a specific time schedule for implementation, while

Government funded schemes can often (but not always) be more flexible with respect to the rate of implementation. The project approach may therefore be defined as "a planned undertaking which is a set of interrelated and co-ordinated activities designed to achieve objectives within a given budget and period of time."

This differs from a national programme where the time schedule is more flexible and the budget provision tends to fluctuate according to Government fiscal policy. At times of exceptionally strict budget constraints, strong Government commitment is required to maintain budget provision in line with project schedules.

OBJECTIVES OF INTEGRATED RURAL DEVELOPMENT

There can be a broad range of objectives for integrated rural development projects, from the very ambitious such as attaining national average income or employment goals, to more moderate aims such as maintaining the size of the local population, or even more modest ones such as simply improving the productivity of one crop.

A reasonable objective, however, for most integrated rural development projects is "to lay out the project interventions in the area economy that are most likely to lead to important improvements in the way the economy operates and in the ability of the area population to achieve a fitting living ..."

COMPONENTS OF INTEGRATED RURAL DEVELOPMENT

The components of integrated rural development projects must vary from one area to another in relation to the specific problems and potential of each area. The following is a list of project components that should be considered but only adopted where appropriate. An integrated approach does not require that all the Components listed should be adopted, but rather that they be considered, and priorities determined because rarely are budget funds available for everything to be done.

Emphasis should be given is the most viable and cost effective components. Integrated rural development projects should consider the inclusion of the following components:

(1) *Income generating sectors:*

- Agriculture, including crop production and animal husbandry and the associated hunting, fishing and forestry.
- Manufacturing industry, including workshops, handicrafts, cottage industry, traditional products and products for which the region is particularly suited. through traffic etc.
- Trade, including the encouragement of markets for local products, serving other areas, sales to
- Tourism, including agro-tourism, special interest and environmental tourism.

(2) *Income generating supporting services:*

- Marketing of agricultural products, processed goods and locally produced manufactures cottage industry.
- Storage for agricultural outputs and inputs.
- Transport of people and goods.
- Supplies of inputs and materials.
- Credit for annual needs, investment and family requirements.

(3) *Research* : There is often inadequate research in lagging areas with respect to both agriculture and natural resources (geological, water and environmental resources). The need for research and exploration programmes should be examined. In the Pitsilia regions there was no research on new vine varieties, while in the Krasokhoria region there was no research on improving productivity on calcarious soils.

(4) *Training* : Training is important not only for farmers through extension services but also for encouraging manufacturing industry and tourist development. The educational system also affects skill levels.

(5) *Labour Force and Personnel Requirements* : The existence of an adequate labour force with the necessary skill requirements is essential. Special provision should be made for the appointment of sufficient state funded personnel, since in many countries procedures for the appointment of state personnel differ from provisions for the Budget (i.e. agreed provisions for teachers, medical staff etc. have to be made).

(6) *Infrastructure* :

- Roads: Access and transportation are among the most serious constraints in mountain areas. Road improvement is likely to be a priority in most cases, especially in mountain regions.

- Utilities: Other infrastructure services, such as water, telephones and electricity are also essential for development and improved living standards, while sewerage can pose problems in mountain villages.
- Farm roads and land consolidation: In most Mediterranean regions and especially in mountain regions, there are serious land tenure problems arising both from the small size of holdings and the fragmentation of the land, as well as access to farmland for vehicles. The most effective way of dealing with this is by land consolidation, which brings the scattered ownership together and provides road access. Alternatively crude farm roads can be established.
- Irrigation: in arid and semi-arid areas irrigation is the major input for improved farm productivity, the productivity of irrigated land being six to twelve times greater than dryland. Mountain areas are generally a source of water for other regions, but their own requirements should not be neglected.

(7) *Social Services* :

- Education: Including pre-school which is important for the labour supply (releasing young mothers for work),primary and secondary. In exceptional cases potential may exist for tertiary level educational institutions which have successfully been used for the development of lagging regions in other countries.
- Health: Provision of medical services, notably doctors, hospitals and medical services (pharmacies, ambulance etc.).
- Welfare: Research is needed on welfare needs (care for old and retarded people, social problems of the very poor etc.).

(8) *Recreation* : The lack of recreation facilities in rural areas is a major factor causing young people to migrate to urban areas.

(9) *Village environment* : The improvement of the village itself and housing is important for future survival. Improvements in village streets, lighting, public buildings, and general environment are important for improving village life.

(10) *Environment* : The environment around the villages and within the region provides opportunities for improvements in income from tourism (including hunting, hill resorts, skiing, agro-tourism, special interest and environmental tourism). Similarly, the built environment such as traditional buildings, also form attractions.

(11) *Institutional Development* : This is important and includes village leadership, creation of credit institutions, co-operatives, management of infrastructure (small irrigation schemes etc.)

GENERAL GUIDING PRINCIPLES

The following general guiding principles should be applied with a view to developing effective and viable projects, that are able to stimulate income generation, improve living conditions and welfare, and contribute to the resolution of problems, in a manner that is likely to be both efficient and self-sustaining.

National policy context: The Project should be in line with national policy. I f this is not the case the project will be continually questioned, the necessary commitment at all levels of Government will not be forthcoming, it will be difficult to justify the additional budget requirements and conflicts of policy both within and between ministries and departments will inevitably arise. Implementation may be facilitated by a clear Council of Ministers decision for project approval including provisions for finance, co-ordination and participation of all ministries involved.

Co-ordination: There must be good co-ordination between the various ministries, departments, state corporations, local authorities, credit institutions and farmers organizations involved. Arrangements for co-ordination should, however, be flexible and subject to the judgement of project management as long as the system works.

Top level support: Project management must have support at a very high levei, either nationally or regionally (i.e. at ministerial or the highest regional level) and should not be under a specialised department.

Control of Project funds: All Project funds should be under Project management and delegated to the other agencies involved, otherwise Project management will lack influence with the agencies involved and co-ordination will become extremely difficult.

Investment criteria: Productive investments should be determined on the basis of the following criteria:

- Comparative advantage.
- Economic viability.
- Self-sustainable investment.

Cost effectiveness: Ali investments should be cost effective, that is economical solutions should be examined both with respect to investment (productive and social)and operating costs (i.e. use of existing buildings and infrastructure where possible, cost sharing etc.).

Subsidies: New subsidies should in general be avoided. Where subsidies are unavoidable they should be in the form of initial incentives and not open-ended, such as income or product price subsidies. Investment subsidies may be considered (i.e. for soil conservation works). Where national subsidies exist the project area should not be discriminated against.

Welfare support: In general production, income or price subsidies should not be provided in order to deal with social problems, because they disrupt economic decisions and lead to other problems (i.e. oversupply, high annual budget costs, high cost in relation to alternatives etc.). Social problems should be dealt with either through investment (if viable or appropriate) or through social programmes such as welfare support which are more selective and should be more cost effective).

User charges: Self-sustained development normally implies that user charges at least cover operating costs, consequently the issue of charges to the community has to be examined. Charges should relate to goods such as water to ensure efficiency in its utilization. The provision of services and investment should be considered within the concept of user charges and national policy, and rural areas should not be discriminated against. Contributions of the community to investment (roads, water, public buildings etc.)are also a form of user charges.

Cost contributions: The contribution of village communities to costs of works such as public buildings, village streets, small irrigation systems etc. ensures community participation in projects and that the scheme has priority for the village.

Community involvement: Communities should be involved in the—project both formally and informally, including project planning, implementation and contribution to costs. This may lead to changes in project design to meet community needs.

Priorities and phasing: Priorities have to be determined from amongst all the various options, consequently same components will inevitably be dropped and assessment criteria determined. The application of priorities and the logical sequencing of schemes determines the composition and time schedule for the project.

Flexible implementation: Rural development projects are complex and .difficult to implement, and even the best prepared projects have to be changed during implementation. Flexibility within the context of the general framework and project objectives should be possible. Some of the most successful projects came close to total failure, but were implemented with good results after changes were made.

ORGANIZATION OF INTEGRATED RURAL DEVELOPMENT

Project Preparation

The lead role in project preparation should be taken by an agency independent of the line ministries, which is capable of co-ordination and directing a large number of other agencies, and which they will accept in this role without worrying about encroachment on their functions. The lead agency should have experience in dealing with multi-sectoral projects and be given the functions, funds and status to ensure co-operation. In many countries the planning ministry or bureau or the regional administration may be best place to undertake the lead role. Alternatively, Project preparation could be undertaken by independent consultants assisted by Government Departments.

The participants in project preparation should be the lead agency, line ministries, regional/district administrations, local authorities, state corporations and non-government organizations (Chambers of Commerce/lndustry, farmers organizations, co-operatives etc.). One method of ensuring smooth co-operation is by appointing representatives of the ministries and agencies, who also act as liaison for their agencies, In this way one person in each organization is responsible for seeing that the obligations of that body are met. Unavoidably, however, major issues will have to be dealt with at the top management level, or in the event of clashes, at the ministerial level.

Consultations with the local authorities and local NGO's is essential, both in informal contacts and formal project preparation. The agreement of local authorities is absolutely essential where they are expected to contribute to costs and where infrastructural programmes run through their administrative boundaries.

The main problems in project preparation involve co-ordination, particularly where there are disputes between agencies as to jurisdiction, and inflexible rules that have not been adjusted to meet the needs of multi-sectoral projects. This is true both of Governments and international agencies.

Project Implementation

There are two basic approaches to implementation of multi-sectoral integrated rural development projects: one is the establishment of a self-contained fully staffed and equipped independent organization (perhaps a regional development authority) and the other is to undertake the project through existing agencies with a small management and co-ordination unit having overall responsibility for the project. Both approaches have their advantages and disadvantages, but in smaller countries the second project management unit approach is preferable because it is more cost effective.

(1) Self-Contained Implementation Unit : This approach can only be justified with respect to very large projects, in big countries, because it essentially entails the recruitment of personnel for all project implementation activity in all sectors. This is in fact virtually impossible because co-ordination with other departments or agencies and local authorities is still required, and no one organization can cover the full scope of activity envisaged in most integrated rural development projects. The integration of such units into the existing administrative and legal structure poses very serious problems, and ministries afraid of encroachment on their jurisdiction may be uncooperative. This type of structure is also costly, and complex since it is unlikely that line agencies and local authorities can be altogether excluded and there may be duplication of activity. The main problem, however, is in ensuring project continuity, maintenance and further development when the project is completed and the special self-contained unit disbanded.

The great, advantage is that much red-tape and effort is made unnecessary and project management has considerable powers to move quickly with project implementation, because the unit's dependence on other organizations is minimal. This approach is suitable where Government administration and technical capability is weak.

(2) Co-ordinating Management Unit : The establishment of a small management co-ordination unit is more cost effective because it involves the recruitment of only a small nucleus of staff around a project manager, with responsibility for ensuring that all the line ministries and other agencies fulfil their commitments to project implementation within the budget allocation and the time schedule for implementation. The main problems arise because the authority of the unit may be challenged. This can be overcome by ensuring that the Project Manager has sufficient status and authority assigned by the Council of Ministers and is independent (under the Prime Minister, Planning Ministry or Regional Authority) or at least at ministerial level. It is essential for the status and control of the management unit that it control all project funds and assign them to line ministries and other agencies. This provides an incentive for good co-ordination.

The great advantage of this approach is that it ensures continuity since implementation is undertaken through existing bodies. It is most suitable, however, where Government administration is strong.

(3) Essential Conditions for Implementation : The essential conditions for the organization of the implementation effort are as follows:

i. Clear authority for project management to do the job of implementation.
ii. Instructions to ministries to cooperage and assign representatives and/or liaison officers.
iii. Control of the project budget by project management.
iv. Priority to be given by ministries to the project components.
v. Appointment of a very good, full-time project manager.

(4) World Sank Evaluation of Integrated Rural development : In 1988 the World Bank undertook an evaluation of the integrated rural development projects that it had financed and supervised. The main conclusions were as follows:

i. The projects "often did not perform well".
ii They were costly.
iii. They were administratively complex.
iv. In South America and Africa (where government structure is relatively weak) governments could not implement the projects effectively.
v. Large projects could be handled only at central government level.
vi. Implementation time schedules were too tight.

The reasons for the poor results were found to be mainly the following:

i. Inappropriate project design. More specifically ethnic diversity was disregarded; there was insufficient knowledge on beneficiaries; imbalances in economic and political power were not taken into account, and there was competition for labour from other sectors.
ii. Prices were either inappropriate (i.e. wrongly fixed by Government) or fails in prices were not predicted.
iii. Inadequate attention was given to marketing.
iv. Insufficient attention was given to land tenure issues, and the fact that no investment is undertaken if the and is not owned by farmers and there is insecurity of tenure.
v. There was insufficient government support, that is a lack of commitment to the project.

(5) Conditions for Success : The experience gained from integrated rural development projects suggests that the conditions for success are as follows:

i. Strong government commitment to the Project.
ii. Effective administration in place at central government, regional and community levels.
iii. Good project preparation.
iv. Clear objectives must be set with respect to short and long term objectives and by component.
v. A good, full-time, project manager must be appointed.
vi. Flexibility in implementation.
vii. Authority for implementation must be given to project management.
viii. Good co-ordination and co-operation by all ministries, departments, agencies, local
ix. government and N G O s is essential.
x. Project management must have overall budget contro!.
xi. The project should be economically viable, at least with respect to its productive components.
xii. The project should be cost efficient.
xiii. Local resource mobilisation is important for self sustainability and development impact.

In addition to the aforementioned it is necessary for the project to bring about a change in the perception of the project area, from disillusionment to and optimistic assessment of future prospects.

This is essential for there to be sufficient investment for self-sustained growth.

(6) The Role of the Integrated Rural Development Approach :Integrated rural development has a role to play in identifying opportunities and local needs, and presenting a framework for raising incomes and improving living standards. This involves complex planning, co-ordination and execution combined with good project management. Such integrated projects facilitate development both through investment and infrastructural works and by changing the perception and the reality in a region, so as to encourage self-sustained growth.

RURALITY AND POVERTY

Most of the world's poor live in rural areas. IFAD estimated in 2001 that among the poorest 1.2 billion people in the world, surviving with less than a dollar per day, three out of four lived in rural areas. They constitute the poorest fifth of world population and do not earn enough to cover their food needs. In a recent World Bank study, Ravallion et al. (2007) estimate that in 2002, 75% of the developing world poor still live in rural areas. Part of this correlation between rurality and poverty is given by the fact that some countries, as we have seen, indirectly define the poor as rural. As one of the most accepted characteristics of development is a secular decline in the share of agriculture, countries with larger rural populations shares are expected to be poorer since the main activity in the rural economies is likely to be agriculture (we explore the depth of this link below). However, the question begs an answer, how much are poverty and rurality correlated? To get a measure of correlation we collect all the available figures of poverty measured as the percentage of population in each country living under 2 Purchasing Power Parity (PPP) US dollars per day. The benefit of this measure of poverty is that it is comparable, because it offers a comparable poverty line which accounts for the differences of purchasing power of 1 dollar in each different country.

AGRICULTURE AND RURAL DEVELOPMENT

To what extent is rural development the same as agricultural development? In other words , in which case does a sectoral (agricultural) driver to rural growth is indispensable? To answer this question we need to have an idea of the importance of agriculture within the rural world.

There is no direct measurement to uncover the importance of agriculture within the rural economy, but we will show alternative roads which indirectly point to that measure. Agriculture, is an important component of most rural economies especially in the developing countries. It was shown above that the size of agriculture within the local economy is sometimes used to define rurality. Therefore, any successful rural development strategy will contain an agricultural development component; but they are not the same thing.

While agricultural development aims at improving the welfare of populations through sustained improvements in the productivity of the agricultural sector, rural development aims at the improvement of welfare of rural populations through the sustained growth of the rural economy, which includes agriculture, but may not be its only component and not necessarily the most dynamic. Ideally, if we wanted to know the importance of agriculture within a rural area, we would look at the GDP or output figure for that region, and measure the share of agricultural output within the total value added of the region. Unfortunately only few countries have aggregate output figures available by region (state or province), and even at this regional level, output is aggregated for both rural and urban areas.

We show above that the importance of agriculture within the national economy falls with development, but this does not necessarily mean that the relative importance of agriculture in the rural economy also falls with development (given that rurality also falls with development). Poorer countries, with lower per capita income, and with higher incidence of poverty, not only are more rural, but in their economies agriculture has a higher relative weight. Thus, the poorer the economy, the more important agriculture is for its rural and overall development. Another road to measure the value of agriculture in the rural economy is to examine the share of income from agriculture to total rural income. This can be done, thanks to household surveys that measure income, most of them implemented in developing countries since the 1950's to understand poverty. Unfortunately, the way researchers have measured rural agricultural income vis-à-vis non-agricultural income (also known as rural non-farm income) has varied too much. Examples of these conceptual inconsistencies are: to add remittances (which in countries like Pakistan and in Central America can amount to more than 5% and even 10% of household incomes) to rural non-farm income, when remittances are not rural (sometimes not even national) income. Other inconsistencies arise with income from wages of agricultural labor. Some authors add wages accrued in farms outside their own to non-farm income, when it is clear that this is agricultural income. Further problems arise when wage income due to the absence of information can not be assigned to any particular sector, in these cases all of wage income can be added to either the farm or non-farm sector. Some authors even add food sales to non farm income. Then there are gray areas, like the way in which to value own agricultural consumption; or income from fishing and forestry that could be added to either farm or non-farm income. These conceptual differences and inconsistencies highlight the need for comparable measures of agricultural and nonagricultural rural income to make further meaningful cross-country comparisons and studies. However, for our purpose of uncovering any links between rural agricultural income and development, these measures, alas inconsistent, will be helpful.

For this study we have collected share of rural agricultural income for 120 country-year combinations. Three country-year combinations are repeated with different figures, and we keep all the information as we are in no condition to discriminate the inaccurate information. The main sources are FAO (1998), and Lanjouw and Feder (2001), that made comparable recollections of rural nonfarm income shares, and our own collection from more than 30 monographs, mostly from rural household survey studies. This database contains information from four continents and it spans the period 1950 to 2002. The results confirm that more developed rural economies have a lower share of agricultural income in total rural income. This relationship is statistically significant, and the estimated elasticity indicates that a 1% growth in rural per capita agricultural income reduces the share of agriculture in total rural income by about 0.12%. Also, the differences in income (development stage) explain one fifth of the variability of the share of agricultural income in total rural income, which is a good fit considering the possible inconsistencies in the measurement of agricultural income; and that geographic, agroecological and climatological differences probably explain a lot of the observed variability.

The negative relationship between rural development and the share of agricultural income can also be confirmed by examining time series data for particular countries. In Japan for example the share of agricultural income in rural income dropped from 82% in 1950 to 30% in 1980; similar drop can be observed in India, 92% to 62% in the period 1958-1994; and in China from 90% to 67% in the period 1980-1997. In Bangladesh the share of agricultural income in rural income actually went up in the period 1963-1982 from 82% to 92%, but real income also dropped in this Asian country during this same period (agricultural GDP per rural person fell from 103 to 91 US$). Thus, the cross country as well as the time series data indicates that a feature of rural development is that the share of agriculture in rural income falls with development. Nonetheless, as Table 6 indicates, agriculture is the main source of income for rural households, responsible for more than three fifths of total rural incomes.

CONCLUSIONS AND SOME POLICY IMPLICATIONS

The last few years have seen an increased attention of the international development community on agriculture and rural development. If the data for the last 3 years constitute a real trend then we are witnessing a real increase in official development assistance to agriculture and rural development. The emergence of private donors and foundations (such as the Gates foundation) with keen interest in transferring resources to agriculture is a welcome development. Major multilateral donors (such as the World Bank) are looking at agriculture as an engine for poverty reduction for most developing countries and regions and a fundamental component of a growth and poverty reduction strategy for the poorest, agriculture based economies. There seems to be increased interest by domestic and foreign private capital (including multinational agro-industrial firms) for investments in sectors up and downstream of production agriculture. The potential of agriculture as a source of bioenergy promises to attract further attention to the potential of the sector to produce and address global food and energy needs.

This Chapter provides analytical evidence and arguments in support of the proposition that agriculture and the rural economy are fundamental for obtaining substantive and sustainable gains in the fight against poverty. Even using the existing, highly imperfect measures of "rurality" (which very likely underestimate the economic importance and dimensions of the rural space and the extent of rural poverty) poverty is intrinsically linked to rurality. What the evidence shows is that even when population movements and demographics reduce the share of the rural in total developing country population, poverty will still be a predominantly rural phenomenon. The experience of Latin America (the most urbanized of the developing regions) is instructive in this regard: even in countries where the number of rural poor are less than the number of urban poor, the poverty rate is nevertheless higher in the rural areas. It also shows that the analytical underpinnings of a development strategy with a strong anti-agricultural bias do not hold in light of new data and evidence. Agriculture has strong links with other sectors in many countries and, a productivity-induced agricultural expansion can "pull" other sectors with it, increase economic activity and employment opportunities in the rural areas. The paper also shows that, while a regional or spatial approach to the development of the rural economy is a reasonable proposition for countries with good infrastructure, functioning labor and other markets, there is substantial scope for a sectoral (agricultural) entry point for rural and possibly overall development and poverty reduction. This proposition is further corroborated by the results of Davis et al. (2007), who show that despite income diversification by rural households; households in the lowest expenditure categories have a larger share of agricultural in their total income than households in higher income groups.

The balance of entry points for rural development and poverty reduction is bound to change, as food systems change both globally and, especially in developing countries. However, such a change hides both opportunities and threats. First, the role of agriculture in the structure of the economy will decline in the process of development. However, the data on the agricultural transformation shows that the labor share of agriculture declines much slower than the share of agriculture in national GDP. Unless policies and investments are put in place to foster agricultural productivity, there is a danger that the decline of agriculture will be accompanied by increased rural poverty some of which will find its way into the urban areas. At the same time, policies and programs which increase the human capital of the rural poor and allow them to enter a more remunerative labor markets are powerful tools to ensure a smooth transition of people out of agriculture without increasing poverty. At the same time a transformation of the agricultural sector itself is underway. Growing per capita incomes have increased and will increase further the demand for high value food products and for the quality and safety characteristics of foods. Such shifts in demand are reflected in the structure of the food systems even in some of the poorest developing countries towards the spread of the modern food chains and consolidation of the production, distribution and retail segments of food markets. The role of the more traditional "chain" will shrink overtime. Therefore, while increasing productivity for food staples oriented agriculture will still be an important anti-poverty entry point for many poor areas in the world, the high requirements in terms of skills and capital by the "new agriculture" point to the need for increased access to modern technologies (research and extension systems) which promote the higher quality and safety standards demanded by consumers.

Will trends towards (domestic) market integration, and globalization affect the strength of the links between primary agriculture and up-stream and down-stream sectors? On one and, market integration will sever the links between agriculture and the local economy, but will open new links to larger markets (regional, national or even global). Opening the agricultural sector to foreign but also domestic competition will challenge the position of smallholders in national markets. The non-farm rural economy will also be subjected to competition pressures from cheaper consumer goods distributed by supermarkets and similar retail outlets as they expand to the rural areas of developing countries. Small scale and artisanal food processing may give way to more organized, capital intensive processing plants and a similar scenario may be thought for other activities up or down-stream primary agriculture (such as transport or input processing, etc.). The speed and extent to which these changes will occur will vary by context as will vary the net effects on employment etc. (Reardon et al. (2007), Stamoulis et al. (2004)). In such cases, even if the "inverse" relationship between size and productivity were to hold in primary production (commodity production) especially of staples, larger farmers will more than likely have an advantage in producing what will be increasingly demanded: high value products requiring capital intensive technologies and human and managerial capital.

However, the future of smallholders and rural livelihoods is not without opportunities: access to larger markets and higher value alternatives will also be available for those who innovate and are able to take up opportunities presented by the changing system. Public policy will have a significant role to play in this context. In addition to providing the "traditional" public goods it also needs to provide assistance to farmers in dealing with new product requirements, and creating the proper institutional and regulatory framework to enable smallholders to organize so as to exploit available economies of scale and promote competition.

RURAL DEVELOPMENT ACTIONS

Rural development actions are mainly and mostly to development mostiy aim the social and economic development of the areas. In Europe, rural development actions also integrate environmental management as a core component. Rural development programs are usually top-down from the local or regional authorities, regional development agencies, NGOs, national governments or international development organizations. But then, local populations can also bring about endogenous initiatives for development. The term is not limited to the issues for developing countries. In fact many of the developed countries have very active rural development programs. The main aim of the rural government policy is to develop the undeveloped villages. Rural development aims at finding the ways to improve the rural lives with participation of the rural people themselves so as to meet the required need of the rural area. The outsider may not understand the setting, culture, language and other things prevalent in the local area. As such, general people themselves have to participate in their sustainable rural development. In developing countries like Nepal, India, integrated development approaches are being followed up. In the context of many approaches and ideas have been developed and followed up, for instance, Bottom-up approach, PRA-Participatory Rural Appraisal, RRA- Rapid Rural Appraisal etc.

Participatory Rural Appraisal (PRA)

Participatory rural appraisal (PRA) is an approach used by non-governmental organizations (NGOs) and other agencies involved in international development. The approach aims to incorporate the knowledge and opinions of rural people in the planning and management of development projects and programmes. Hundreds of participatory techniques and tools have been described in a variety of books and newsletters, or taught at training courses around the world. These techniques can be divided into four categories:

§ Group dynamics, e.g. learning contracts, role reversals, feedback sessions

§ Sampling, e.g. transect walks, wealth ranking, social mapping

§ Interviewing, e.g. focus group discussions, semi-structured interviews, triangulation

§ Visualization e.g. venn diagrams, matrix scoring, timelines

To ensure that people are not excluded from participation, these techniques avoid writing wherever possible, relying instead on the tools of oral communication like pictures, symbols, physical objects and group memory. Efforts are made in many projects, however, to build a bridge to formal literacy; for example by teaching people how to sign their names or recognize their signatures.

RURAL DEVELOPMENT IN INDIA

The Rural Development in India is one of the most important factors for the growth of the Indian economy. India is primarily an agriculture-based country. Agriculture contributes nearly one-fifth of the gross domestic product in India. In order to increase the growth of agriculture, the Government has planned several programs pertaining to Rural Development in India.

The Ministry of Rural Development in India is the apex body for formulating policies, regulations and acts pertaining to the development of the rural sector. Agriculture, handicrafts, fisheries, poultry, and diary are the primary contributors to the rural business and economy.

The introduction of Bharat Nirman, a project set about by the Government of India in collaboration with the State Governments and the Panchayat Raj Institutions is a major step towards the improvement of the rural sector. The National Rural Employment Guarantee Act 2005 was introduced by the Ministry of Rural Development, for improving the living conditions and its sustenance in the rural sector of India.

Rural Development in India-Schemes

Pradhan Mantri Gram Sadak Yojana (PMGSY): This is a scheme launched and fully sponsored by the Central Government of India. The main objective of the scheme is to connect all the habitations with more than 500 individuals residing there, in the rural areas by the means of weatherproof paved roads.

Swarnjayanti Gram Swarozgar Yojana (SGSY): This was implemented as a total package with all the characteristics of self employment such as proper training, development of infrastructure, planning of activities, financial aid, credit from banks, organizing self help groups, and subsidies.

Sampoorna Gramin Rozgar Yojana (SGRY): This scheme aims at increasing the food protection by the means of wage employment in the rural areas which are affected by the calamities after the appraisal of the state government and the appraisal is accepted by the Ministry of Agriculture.

Indira Awaas Yojana (Rural Housing): This scheme puts emphasis on providing housing benefits all over the rural areas in the country.

Rural Development in India-Organizations

Department of Rural Development in India: This department provides services such as training and research facilities, human resource development, functional assistance to the DRDA, oversees the execution of projects and schemes.

Haryana State Cooperative Apex Bank Limited: The main purpose of the Haryana State Cooperative Apex Bank Limited is to financially assist the artisans in the rural areas, farmers and agrarian unskilled labor, small and big rural entrepreneurs of Haryana.

National Bank for Agriculture and Rural Development: The main purpose of the National Bank for Agriculture and Rural Development is to provide credit for the development of handicrafts, agriculture, small scaled industries, village industries, rural crafts, cottage industries, and other related economic operations in the rural sector.

Sindhanur Urban Souharda Co-operative Bank: The main purpose of the Sindhanur Urban Souharda Co-operative Bank is to provide financial support to the rural sector.

Rural Business Hubs (RBH): RBH was set up with the purpose of developing agriculture. The Rural Business Hubs Core Groups helps in the smooth functioning of the Rural Business Hubs.

Council for Advancement of People's Action and Rural Technology (CAPART): The main purpose of this organization is to promote and organize the joint venture, which is emerging between the Government of India and the voluntary organizations pertaining to the development of the rural sector.

✱✱✱

3

RAPID RULER APPRAISAL

INTRODUCTION

Unfortunately, there is no generally accepted definition of RRA. RRA is more commonly described as a systematic but semi-structured activity out in the field by a multidisciplinary team and is designed to obtain new information and to formulate new hypotheses about rural life. A central characteristic of RRA is that its research teams are multidisciplinary.

Beyond that, the distinction between RRA and other research methodologies dependents upon its multidisciplinary approach and the particular combination of tools that in employs. A core concept of RRA is that research should be carried out not by individuals, but by a team comprised of members drawn from a variety of appropriate disciplines. Such teams are intended to be comprised of some members with relevant technical backgrounds and others with social science skills, including marketing research skills. In this way, it is thought that the varying perspectives of RRA research team members will provide a more balanced picture. The techniques of RRA include:

- interview and question design techniques for individual, household and key informant interviews
- methods of cross-checking information from different sources

- sampling techniques that can be adapted to a particular objective
- methods of obtaining quantitative data in a short time frame
- group interview techniques, including focus-group interviewing
- methods of direct observation at site level, and
- use of secondary data sources.

McCracken et al. describe, rather than define, RRA as an approach for conducting action-oriented research in developing countries.

Ellman offers a good example to compare the "traditional" rural development research to RRA. He was requested to analyse the achievements of four types of land settlement schemes in Sri Lanka and to recommend a policy for future settlement in one million acres of cultivable land. He was advised, by a statistician, to sample 20% of the settlements in operation. This meant 80 settlements and two and a half years of field survey were carried out by two experienced researchers. Finally, he and his colleague, helped by two assistants, conducted "full social, economic and attitudinal surveys" with a sample of 20 settlements. Nine months were needed to collect the data, and six months to write it up. The result: a two-volume report of 305 pages. By the time the message (Ellman says "quite a simple one") was absorbed by those who needed to know it, the government had changed and suddenly the study had lost much of its relevance. With the idea of RRA in mind, he is "convinced that the same message could have been put across more quickly, cheaply and effectively, with evidence drawn from a smaller, purposively selected and studied sample and with no significant reduction in reliability".

Ellman was later commissioned to carry out another study, and having learnt from that earlier experience, he identified the minimum amount of data that was required and was likely to be effectively used for planning purposes. A team of ten people collected data in six weeks, and produced a 25-page report with clear recommendations, which were broadly accepted by the government and the international agency. The project, concerning integrated rural development, was finally implemented.

The application of RRA has been quite wide as regards rural development, for example in health, nutrition, emergencies and disasters, non-formal education, agroforestry, natural resource assessment and sociology approaches. RRA has also been applied in agricultural marketing, although the marketing orientation of RRA studies has not been very well defined.

Thus the term rapid appraisal does not refer to a single technique but to a range of investigation procedures. Their chief characteristics are that they take only a short time to complete, tend to be relatively cheap to carry out and make use of more 'informal' data collection procedures. The techniques rely primarily on expert observation coupled with semi-structured interviewing of farmers, local leaders and officials. In substance the techniques of RRA have much in common with the social anthropologist's case study approach but are executed over a period of weeks, or at most months, rather than extending over several years. To date RRA has mainly been used in the field of rural development as a short cut method to be employed at the feasibility stage of project planning.

Rapid rural appraisal is a set of techniques that can be applied as a preliminary stage when embarking on surveys of farmers. The technique essentially involves an informal, rapid, exploratory study of a specified geographical area designed to establish an 'understanding' of local agricultural conditions, problems and characteristics. They can provide basic information on the feasibility of beginning a survey project in an area, particularly when one is intending to survey an area about which little is known.

DIFFERENCES BETWEEN RAPID RURAL APPRAISAL AND OTHER APPROACHES

Chambers describes the orientation of RRA as a "fairly-quick and-fairly-clean" appraisal, and as opposed to the fast and careless studies (he calls them "quick-and-dirty" studies) and the slow and excessively accurate approaches ("long-and-dirty").

On the one hand, the most common form of fast, rough studies - the "quick-and-dirty" approach - is termed rural development tourism, that is the brief rural visit by the urban based "expert". Although Chambers says that it can be cost-effective for the outstanding individual, it is important to be clear that it can be actually low-cost research, but of course not cost-effective. This practitioner, as opposed to the formal academic, wants quick insights and quick results. "Brief rural visits, snatches of information here and there and a few observations, anecdotes and impressions are put together as the basis for time-bound judgements and decisions". Commonly rural development tourism means highly biased results, especially concerning anti-poverty biases. In other words, it has large misperceptions and misinterpretations of the rural reality, and not only due to lack of time to carry out a deeper research.

Chambers cites the main biases of rural development tourism as being: spatial (urban-tarmac-and-roadside biases, that is going only to easily accessible places), project (neglecting non-project areas), personal contact (meeting the less poor and more powerful rural people, men rather than women, users of services rather than non-users, and so on), dry season (travelling in the post-harvest or post-rainy season, when it is easier) and politeness-and-protocol bias (lack of courtesy and convention, lack of adaptation to local conditions, shortage of time, etc.).

On the other hand, formal and accurate studies - the "long-and-dirty" ones Chambers goes on to describe - are longer and more costly solutions preferred by "well-trained" professionals. They are preferred by the academic community, interested more in detail, precise observation and measurement and rigorous and respectable methodology and with a generally rather unhurried concern for knowledge for its own sake in the longer term. Formal studies do not generate information in the early stages, and some, though not all of them, are academically excellent but finally useless, very likely because of the lack of opportunity to induce the desired effects or results. Many are never used: never coded, punched, processed, printed out, examined, written up, read, understood or known to actually change action. Moreover, many studies rarely communicate the knowledge gained; researchers write in a way that makes it unavailable or unintelligible to bureaucrats in the formal planning system. Chambers says "... rural surveys must be one of the most inefficient industries in the world...; the longer the research takes, the longer and less usable the report tends to be and the greater the time available for sweeping the dirt under the carpet". The biggest single blockage is the failure to "treat statistics as servant rather than master". Cases to illustrate the fact can be seen all over the world: papers that, if finally written, are too late to induce the change desired (for instance, the case already described in Ellman's experience). McCracken et al. summarised the main arguments against inefficient formal (traditional) studies demonstrating that these conventional methods have a long duration, fixed and formal structure, limited scope, weak integration, exhaustive depth, "top-down" direction, low integration with local farmers, high cost and inefficient use of time and manpower.

As stated earlier, the point is to determine which information is really relevant, opportune, understandable and actually useful for the decision makers: the middle zone (between both extremes, the short-and-inaccurate and the long-and-excessively-accurate) of greater cost-effectiveness. The task is obviously not easy and requires experience, knowledge, and perhaps, as it has been even suggested to define RRA, a lot of "common sense". It is therefore important to realise that RRA procedures are more and not less demanding of expertise, when compared to the "dirty-short" and "dirty-long" approaches. Optimal ignorance can only be achieved if investigators are both well informed and sensitive to what they may not know.

TEAM COMPOSITION

One of the main characteristics of RRA is to work with a multidisciplinary team. The advantages have been already discussed. Chambers is emphatic in pointing out that the argument that it is necessary to have an integrated and coordinated approach to research cannot be used as an argument for having only one well-informed and intelligent person to do it all. Moreover, Beebe proposes not using the term Rapid Appraisal to describe studies done by one person.

The question remaining is, "which disciplines have to participate?" The point does not seem to be very critical, since for example Shanner et al. think that the disciplinary speciality of team members is not critical as long as "several" disciplines are represented. For agricultural marketing purposes, and of course depending on the objectives of the research, the best solution seems to be to have at least a team of two researchers, one with natural-sciences background (agriculture and related disciplines), and another with social-sciences background (e.g. economist, sociologist), but both with marketing knowledge. This background of course is not definitive, yet the recommendation is still to have at least two researchers. The particular skills of researchers are discussed in the next chapter.

The recommendation is to recruit both men and women to be included on the research team, to overcome the difficulties and take advantage of the situations associated with researcher gender, as well as to provide possible different insights.

Another recommendation is for researchers to have "some" familiarity with aspects of the systems being investigated. Teams should be composed of a mix of insiders and outsiders. Insiders or people very familiar with the area will provide a high-knowledge perspective to the problem. The outsider's participation may be extremely valuable to the insiders in identifying possible options and in noting constraints that might otherwise be overlooked. Outsiders also can gain insights and knowledge that can guide their research activities away from the farm. These considerations are in addition to the points discussed above concerning participation of farmers themselves.

To carry out research with real interaction and participation of all members of the team it is recommended to work with smaller teams rather than larger teams; a ten-member team is probably too large. For example, large teams working in the same interview simultaneously can intimidate rural people.

Opinions differ on how to structure the time of RRMA, but there is almost universal agreement on the importance of dividing time between collecting data and team interactions to make sense out of the collected data. Interactions between researchers at the end of each day and at the end of the field work seem to be essential in determining the success of the RRMA. Scheduling RRMA time can ensure that time for group interaction will be adequate and that a variety of different activities can be covered in a short period of time. As an initial phase, that is even before deciding on how to structure the plan, it is necessary to decide on whether to invest in a preliminary visit by one or two members of the team, expecting them to explain the forthcoming research, find a place to work (for the team sessions), arrange vehicles, identify local participants, request for information, and so on.

The rapid rural appraisal technique is straight - forward to administer but can be physically demanding. It cannot be accomplished simply by driving along a main road looking at fields (although such a method may be a way to begin). The successful appraisal may require tracking over fields in high temperatures and/or over difficult terrain.

Researchers must be prepared to collect information in the fields, market places, or wherever farmers' daily routines take them.

The rapid rural appraisal requires mental and methodological flexibility. It does not proceed like the 'formal questionnaire survey' where predetermined hypotheses are tested. Instead, important questions, issues and the direction of the study emerge as information is collected. This is not to say that the informal survey lacks logic, but that one must be able to accommodate new information and adjust research plans accordingly.

THE PRINCIPLES OF RAPID RURAL APPRAISALS

The following are the principles of RRA agreed by its practitioners, in spite of the fact that there are different opinions and criteria concerning them. These are general principles of theory:

1. Optimising trade-offs: relating the costs of learning to the useful truth of information, having tradeoffs between quantity, relevance, accuracy and timeliness of the information acquired, as well as its actual use. Trade-offs in this sense are not merely mathematical ratios, but they also entail, in the context of cost-effectiveness, alertness, observation, imagination and the ability to pursue serendipity.
2. Offsetting biases: through introspection, it is necessary to identify cognitive biases and deliberately offset those biases. The recommendations are: to be relaxed and not rushed; listening not lecturing; probing instead of passing onto the next topic; being unimposing instead of imposing; and seeking out the poorer people and what concerns them.
3. Triangulating: using more than one technique/source of information to cross-check answers, that is comparing and complementing information from different sources or gathered in different ways. It also involves having team - multidisciplinary - members with the ability to approach the same piece of information or the same question from different perspectives.
4. Learning from and with the rural people: this means learning directly, on-site, and face-to-face, gaining from indigenous

physical, technical, and social knowledge. Farmers' perceptions and understanding of resource situations and problems are important to learn and comprehend because solutions must be viable and acceptable in the local context, and because local inhabitants possess extensive knowledge about their resource setting.

5. Learning rapidly and progressively: this means the process of learning with conscious exploration, flexible use of methods, opportunism, improvisation, iteration, and cross-checking, not following a blueprint programme but adapting through the learning process. However, this could sound again as a non-systematic way of carrying out research.

A fundamental principle is the making of contact with the rural population in a learning process. This aspect must be one of the focal points.

AGRICULTURAL ISSUES WORTH INVESTIGATION THROUGH APPRAISALS

Before embarking on any appraisal, the research team needs to decide what type of information is needed. The specific objectives of the appraisal should be very clear and a list of important topics for investigation should be drawn up.

In conducting surveys of farmers to establish either demand for, or the acceptability of, new equipment types, the researchers should be looking for information concerning agricultural, farm power and socio-economic issues. Information should be collected oh any related issues in these areas that might affect interest in the equipment being considered. Examples of the types of information that might be collected are outlined below:

Agricultural Issues: for RRA

- Main crops grown (by season)
- Land use intensity
- Extent of inter-cropping
- Soil types and soil conditions (degree of salinity, water-logging)
- Farm sizes areas owned and areas cultivated

- Extent of farm fragmentation; size of individual plots, average number of plots owned, and distances from farmstead
- Ground topography: upland/lowland; land slope, field characteristics (average size, obstacles for efficient use of machinery)
- Irrigation supply canal, rainfed, tubewell, time, day and duration of water received
- Crop yields
- Agricultural practices: methods used and timing for:
- Land preparation
- Planting, sowing, transplantation
- Fertilizer application
- Weeding
- Irrigation, drainage
- Harvesting
- Crop processing
- Transportation

Farm Power Issues: for RRA

- Inventory/ownership of equipment (tractors, implements, tools)
- Use of different power sources for each operation (mechanised, animal or hand)
- Problems experienced in conducting different operations (e.g. labour supply, cost, crop quality, conditions)
- Use of hiring/contracting services for different operations (reasons)
- Labour utilisation type (family, hired), labour cost and availability
- Sources of funds used to purchase machines/equipment in past
- Machine requirement priorities.

Socio-economic Issues: for RRA

- Land tenure systems: tenants, owners, owners-cum-tenants
- Decision making units; extent and influence of social networks on machine/implement/tool purchasing
- Farm incomes: sources and amount

- Availability of resources: sources of funds if required (personal, friends, relatives, agricultural banks, money lenders)
- Age and level of education.

Recording Rapid Rural Appraisal Data

Although data collected through a rapid appraisal is usually for the purposes of one planned survey, it is imperative that the information is recorded in a form which will be useful to subsequent surveys in the longer term. The data should be recorded in such a way that a data-base can be constructed for use as reference material for all future surveys.

The results of appraisal field research conducted by each team for each area should be fully written up in a common format. A useful format would be to construct 'data sheets' on which comments are recorded under the headings outlined earlier.

The results of each team's appraisal maps and data sheets should be combined into master sheets' to enable subsequent ease of reference and storage. Successive appraisal maps should use common scales, keys, symbols, colours, etc. making comparison between two areas and two appraisals possible.

ROLE OF RAPID RURAL MARKETING APPRAISAL (RRMA)

RRA is also useful for supporting decisions towards the improvement of agricultural marketing systems in developing countries. The role that Rapid Rural Marketing Appraisal (RRMA) can play in this broad sense of marketing research lies in the identification and prioritisation of marketing problems, and the evaluation of practical means of improving marketing functions, to meet the needs for expansion coupled with higher performance. The first step is to describe accurately and meaningfully the systems that exist. The next step is the evaluation of structures and performance, and the major forces responsible for changing their relationships. In a broader scope, research must analyse the changing strategic role of marketing in the development process.

A checklist of some general and more specific research activities and topics is suggested in the following, where RRMA seems to be a suitable approach:

- Description, problems of assessment and evaluation of marketing systems for design of improvement projects in production, storage, handling, processing, transportation and distribution, as well as wholesaling and retailing, analysing alternatives
- Analysis of marketing feasibility and social acceptability of performance improvements. Prediction of effects on conduct and structure, as well as social effects
- Monitoring and evaluation of marketing activities
- Design and selection of topics to conduct further or other types of research or trials
- Rural communications design for interventions in nutrition, technology adoption, health, housing, standard of living, education, etc.
- Extension techniques assessment for training and technical assistance
- Design of participatory schemes, like farmer-managed marketing enterprises supported by a development or governmental agency
- Rural organisations assessment
- Understanding of resources use and dependency patterns of local communities
- Design of strategies for reduction of losses and deterioration in quality, and quality improvement of agricultural products
- Design of strategies for reduction in production and marketing activities costs
- Understanding and improvement of price efficiencies
- Evaluation of proposed changes in agricultural marketing systems, in terms of objectives of an efficiency for producers, consumers and intermediaries
- Measurement of performance of the "macro" level in marketing systems, in terms of development goals, such as improvement in nutritional status, proportion of the income spent on food, or increased rural income and standard of living
- Managerial, technical and marketing expertise necessity assessment
- Identification and analysis of existing and potential markets for new food products and processes

- Understanding of the technical, social, and economic constraints of traditional farming systems to define research that will lead to changes of benefit to the farmer
- Development of projects, products and processes having an ex-ante market evaluation
- Facilitation of the commercialisation of socially desirable food products
- Development of pricing, promotion and distribution techniques, using profitability and economic criteria
- Analysis of farmers' perception of farm household risk and uncertainty over practices of production, storage, and distribution, perception of price incentives and other stimuli
- Evaluation of systems using a global perspective: economic requirements of the system, appropriate social values, the interacting institutions, and the flow of final results.

RRMA is also a suitable approach in the commercial sense of marketing and not merely social, that is the research of consumer needs and wants from the viewpoint which looks at rural people as a target market. As in the case of social marketing, the application of RRMA will depend on particular circumstances and research objectives. Some research topics, both general and more particular, to carry out RRMA research having rural communities as target markets are suggested in the following:

- Identification of the needs of communities in rural areas, and of the future direction and strategy of food technology research to satisfy those needs
- Study of the market to discover a target market or a market segment for which producers might develop a product
- To test that a project is "on track", that is according to the wants and needs of the consumer or the end-user
- Appropriate technology design; suitability, acceptability, adoption patterns and impact evaluation of technology changes Identification of intensity and variety of forms taken by consumers' demands
- Identification of demographic and psychographic characteristics and constraints on consumer behaviour that determine and affect the market performance, e.g. economic status, income, life style, consumption habits

- Knowledge of what consumers would like to do, and what they are able to do
- Identification of stratification criteria within rural communities for market segmentation strategies
- Study of rural people's attitudes and cultural characteristics towards nutritional improvements, selection of target groups and development of product to be supplemented, and selection or design of the delivery system
- Study of individual and organisational decision-making processes concerning purchasing and consumption patterns; leadership impact
- Study of consumer needs by analysing attitudes, motivation, and behaviour, e.g. risk perception, price perception, brand differentiation and consumer's knowledge of the quality and range of goods and services
- Development and selection of product concepts, products testing, experimentation on developing the prototype product, specification of production process and products
- Definition of research priorities, determination of variables to analyse, setting up experimentation criteria, definition of hypotheses
- Design, evaluation and monitoring of nutrition programmes
- Development of advertising strategies: definition, development and testing, and monitoring and evaluation; audience considerations, valuation of performance and effectiveness, support for response measurement
- Development of total propositions for rural requirements: product, packaging, positioning and advertising
- Design of packaging for rural conditions: functional considerations, appeal at the wholesaling and retailer level, packaging screening tests, effect evaluation
- Corporate image research for firms and institutions working in rural communities.

RRMA can be used to quickly obtain basic information specifically to aid the generation of hypotheses and the design of questions for a questionnaire. In particular information can be gathered relating to:

- farming system characteristics (e.g. farm sizes and types)

- farmers' decision-making processes (e.g. how decisions are made concerning the purchase of new equipment, and the criteria considered important)
- issues of concern to farmers (e.g. their major farming problems)
- armers' equipment ownership (e.g. which type of farmers own which type of equipment, and where they are located).

With a rapid rural appraisal the researcher acts like an explorer, making a brief survey of the horizon before plunging into the depths of the research from which the wider view is no longer possible. If the researcher observes keenly at the start, the remainder of the survey process stands a better chance of success and time will not be wasted. Without a rapid rural appraisal a researcher may find himself surveying the wrong area, collecting the wrong type of information, asking the wrong people, and precious time and funds can be lost back-tracking.

The appraisal can produce, at a minimum cost, a rich description of life in the farming community and an understanding of local agricultural characteristics that will be invaluable in ensuring that the right areas and people are surveyed and that appropriate questions are asked.

THE RAPID RURAL MARKETING APPRAISAL REPORT

The entire RRMA research process and particularly the report should be made "transparent" to readers of the research, and that concerns reliability and replicability of findings. It is the responsibility of the researcher to demonstrate how his or her conclusions were derived from the data, and in such a way that someone else can follow and, if necessary, replicate the analysis and achieve the same result or dispute those conclusions. As a result, the researcher has to "defend" the validity of his/her research findings and of the whole process, including the general constraints of the context, and the particular constraints when collecting and analysing the data. All these considerations have to be reflected in the research report.

A framework for the definition of criteria for the characteristics for the validity of the RRMA report might be:

1. Natural History: report of which avenues were followed and which were rejected. Discussion of policy context in

which the research was conceived, the original purpose of the study and the initial design; how these developed, and what factors or findings led to major shifts in direction, and how the report relates to the policy context

2. Data collected and techniques: discussion of whether the kind of data collected is that demanded by the research problem. Evaluation of the data quality. Evaluation of procedures followed for data collection.
3. Analysis: suitability of analysis (rigorous, systematic, comprehensive and sensitive) and illumination of policy problem by the data. Evaluation of data analysis procedures.
4. Validity of links between concepts and indicators: considerations of descriptive validity: events and indicators being really what they are though to be; and considerations of conceptual validity: extent to which the concepts and categories used fit the data.
5. Validity of hypotheses or theories: considerations of theoretical validity; way in which concepts are handled and the "coherence" of the resulting theory.
6. Theory kept to the limits imposed by sampling selection: considerations of external validity. Scope and generalisation of the theory dependent on the samples used. Replicability considerations.

Rapid Rural Marketing Appraisal (RRMA) emphasises the essential role that marketing plays in the rural development process for Third World countries, and stresses that it is necessary to appreciate the role of "good" rural marketing research to reduce the risk in rural marketing decision making. RRMA (as an agricultural marketing research approach) represents the contextual understanding of agricultural marketing systems, and the application of ad hoc techniques - as well as correct adaptation of techniques by offsetting and to some extent overcoming the difficulties of carrying out research in the rural environment of developing countries. However, the utilisation of RRMA has to overcome a good number of misconceptions about how the ideal marketing research has to be, since decision makers feel themselves more confident when supporting the decisions in "cold" figures rather than in actual understanding. Moreover, they feel it is always required to have formal quantitative surveys, motivated more by avoiding the negative results than by the positive expectations.

First of all, RRMA represents a paradigm to understand rural development and its marketing implications: understanding development as the result of:

- The complex interaction of variables and context, having multiple and changing relationships, and
- The necessary involvement and participation of farmers (and rural people in general), since the very research process, perceives research as a mutual learning process, including also the research component in the formulation and monitoring of development projects.

The process implies not only the generation of projects for the improvement of "pure" marketing functions. RRMA is indeed a very powerful tool to understand the final marketing implications of every endeavour; after all, development necessarily depends upon the degree to which the prospective beneficiaries recognise it as useful to them.

RRMA understands marketing systems in all their complexity and final implications. The interaction and sharing of insights by multidisciplinary researchers avoids the biased partial views, by analysing the systems (agro-ecosystems) performance, and understanding that they are the result of the interaction of social and natural elements. RRMA analyses the structure of systems and provides a means of predicting the effects of changes and suggesting improvement actions.

RRMA also considers research as searching for and analysing data in a fast and cost-effective way. Moreover, RRMA does not only consider the requirement of the fair amount and the fair quality of the information to support marketing decisions, but also gives sufficient openness to accept and incorporate unexpected but relevant information in an ongoing process. RRMA links flexible techniques in a coherent form to collect on-site, cross-check, validate and analyse data according to those priorities. Equally important, RRMA considers simplicity, relevance and meaning as basic points in presenting findings and suggesting action.

RRMA takes into account the huge cultural diversity of groups of rural population in developing countries. Secondary data collection, direct observation and semi-structured interviewing are very useful techniques for the full on-site understanding and cultural adaptation required by research, if it is objective. Moreover, the participation of farmers in the research process, as well as the inclusion of local researchers and researchers with knowledge about the site and having multiple disciplines, all contribute towards overcoming the difficulties in adaptation of the research to the local conditions, i.e. the adaptation to cultural values and beliefs, language, perception patterns, productive activities, economies, traditions, religion, ethical features, politics, and so on.

RRMA offers a reasonable contribution towards overcoming the difficulties of sampling in a rural environment. It avoids the biases of "rural-tourism" research, and tries to study the events on-site, just where they occur or where the evidence is. As a result, RRMA information, limited obviously by the scope of the investigation, generates a purposive sampling frame of key and casual informants and geographical unbiased transects, without strict statistical representativeness, but with high natural-objective representativeness. This representativeness is enhanced by the triangulation procedures, resulting in increased validity.

RRMA proposes gaining accessibility to rural people by involving them in the research process, that is in learning, collecting, finding and analysing information in open discussions and total interaction. As a result, this participatory research also increases accessibility to sites. In addition, the use of key indicators and the information provided by key informants allows access to information otherwise hard to gather. Other techniques, and especially the case of aerial photography, provide a means of gathering information otherwise not accessible, or difficult to appreciate in all its magnitude.

RRMA is an emerging approach, evolving and improving, and therefore it is still limited. There are types of research than RRMA cannot do, for example that requiring statistical reliability, strict replicability and quantitatively precise conclusions. However, it seems to be always useful in exploring, complementing, supplementing and validating other types of marketing research - that is to say more formal types - in rural environments. Its usefulness will depend on the particular interests of each piece of research, but it is always a viable alternative.

✱✱✱

4

Changing Role and Objectives of Rural Development

INTRODUCTION

The definition of rural development has evolved through time as a result of changes in the perceived mechanisms and / or goals of development. A reasonable definition of rural development would be: development that benefits rural populations; where development is understood as the sustained improvement of the population's standards of living or welfare. This definition of rural development, however, has to be further qualified. In the 1960's and early 1970's the consensus was that intense industrialization was the main characteristic of the perceived development path.

In this context it seemed natural to define rural development as precisely leading into that path: "Rural development is essentially a part of structural transformation characterized by diversification of the economy away from agriculture. This process is facilitated by rapid agricultural growth, at least initially, but leads ultimately to a significant decline in the share of agriculture to total employment and output and in the proportion of rural population to total population. (Johnston, 1970)". Later during the 70's, mostly based on equity considerations, the focus and definition of rural development turned to the provision of social services to the rural poor. This shift was partially founded on the recognition that even under rapid growth of income in rural areas, the availability or equitable access to social services and amenities was not guaranteed.

Rural development may be defined as the development of regions excluding the urban areas such as the towns and cities. Smaller settlements such as villages, farmsteads, and market towns are normally included within the concept of rural, while most of the land area is expected to be used as agricultural land, forest or in its natural state. The concept of rural development has changed significantly during the last 3 decades.

Rural development in general denotes economic development and community development actions and initiatives taken to improve the standard of living in non-urban neighbourhoods, remote villages and the countryside. Economic activities typically relate to the primary sector production and processing of food stuffs and raw materials.

Until the 1970s, rural development was synonymous with agricultural development and, hence, focused on increasing agricultural production. This focus seems to have been driven primarily by the interests of industrialization to extract surpluses from the agriculture sector to reinforce industrialization.

With the focus on increasing agricultural production, the stated objective of most countries was to promote smallholder agriculture. Over time, this smallholder agriculture-centric concept of rural development underwent changes. By the early 1980s, according to Harris, the World Bank defined it as "...a strategy designed to improve the economic and social life of a specific group of people—the rural poor." Four major factors appear to have influenced the change: increased concerns about the persistent and deepening of rural poverty; changing views on the meaning of the concept of development itself; emergence of a more diversified rural economy in which rural nonfarm enterprises play an increasingly important role; and increased recognition of the importance of reducing the nonincome dimensions of poverty to achieve sustainable improvements in the socioeconomic well-being of the poor. The establishment of the Millennium

Development Goals has significantly reinforced the concerns about nonincome poverty. With the paradigm shifts in economic development from growth to broadly defined "development," the concept of rural development has begun to be used in a broader sense. It is also more specific, as Harris noted "in the sense that it focuses (in its rhetoric, and in principle) particularly on poverty and inequality." In more recent years, increased concerns on the environmental aspects of economic growth have also influenced the changes.

Today's concept of rural development is fundamentally different from that used about 3 or 4 decades ago. The concept now encompasses "concerns that go well beyond improvements in growth, income, and output. The concerns include an assessment of changes in the quality of life, broadly defined to include improvement in health and nutrition, education, environmentally safe living conditions, and reduction in gender and income inequalities."

Today there seems to be a universal consensus that the ultimate objective of rural development is to improve the quality of life of rural people. This makes it essential to go beyond the income-related factors such as prices, production, and productivity to a range of non-income factors that influence quality of life and hence inclusiveness of rural development.

Lacroix (1985) exemplifies this line of thought when he explains the difference between agricultural and rural development: "Agricultural Development generally tries to raise agricultural production and productivity and is of a technical nature. It is similar to other efforts to develop physical capital as a means for economic growth... Rural Development, though, by definition is oriented more toward benefiting primarily the poor... Thus, the fundamental distinction between pure agricultural and rural development is the emphasis on capital development for the former, and human capital development for the latter."

Since the 1970's rural development as a concept has been highly associated with the promotion of standards of living and as a precondition for reducing rural poverty. This propoor bias was born from the understanding that, particularly in societies where wealth is extremely concentrated, mean incomes could grow without improving the well being of the most dispossessed. Thus, if the general definition of rural development is accepted, i.e. the improvement of the welfare of all members of the rural populations, then this pro-poor bias is justified.

On the other hand, the focus on human capital formation, through the provision of social services in rural areas has been constantly stressed since the 1970's. Originally, this focus stemmed from social equity considerations: it is fair that all of society's members have access to services like education and health. However, the development of endogenous growth theory in the late 1980's provided macro-foundations for this priority, as this theory proved how permanent growth / development is possible (even in the presence of constant returns to scale) when there is balanced investment in both human and physical capital at the same time.

Inclusive rural development is a more specific concept than the concept of rural development. In broad terms, inclusive rural development is about improving the quality of life of all members of rural society More specifically, inclusive rural development covers three different but interrelated dimensions. The first is the economic dimension that encompasses providing both capacity and opportunities for the poor and low-income rural households in particular to benefit from the economic growth process in such a way that their average incomes grow at a higher rate than the growth of average incomes in the sector as a whole. The economic dimension also includes measures to reduce intra- and inter-sectoral income inequalities to reasonable levels. Second is the social dimension of supporting social development of poor and low-income households and disadvantaged groups, eliminating inequalities in social indicators, promoting gender equality and women's empowerment, and providing social safety nets for vulnerable groups. Third is the political dimension of improving opportunities for the poor and low-income people in rural areas, including women and ethnic minorities, to effectively and equally participate in the political processes at the village level and beyond compared with any other categories of the population within and outside rural areas.

DEFINE RURAL

Having defined rural development it is essential to define what is rural. Unfortunately there does not exist a single methodology, much less a single definition of what constitutes rural. The problem is that patterns of spatial occupation are, inter alia, culturally and historically determined and vary among regions of the world. A natural definition of rurality is to define it by exclusion, as that which is not urban, where urban is defined on the basis of population agglomerations.

In practice there are two main methodologies to define rural. The first methodology is to use a geopolitical definition. First, urban is defined by law as all of the state, region, and district capitals (centers), and by exclusion all the rest is defined as rural. Countries like Colombia, El Salvador, Dominican Republic, and Paraguay follow this methodology. In all of these countries urban population is defined as that living within the "cabecera municipal" the municipality's head or center. The drawbacks of this methodology are obvious: populations that live outside the geopolitical limit of a city (specially in a growing city) are miscounted as rural; while population living in tiny municipalities in sparsely populated regions is miscounted as urban.

The other popular methodology is to use observed population agglomeration to define urban. In this case populations that live within an area where contiguous households form populations larger than, say 2,000 inhabitants are considered urban, while by exclusion the rest is defined as rural. This methodology seems more attractive because it establishes a clear threshold; unfortunately this threshold varies widely around the world. In countries like Uganda, an agglomeration of only 100 inhabitants constitutes an urban settlement, while in countries like Nigeria and Mauritius the minimum agglomeration for urban areas is 20,000, and even 30,000 in Japan. In spite of these large variations certain thresholds are popular, like 2,000 in Chile, Argentina, Bolivia, Israel, and France; 2,500 Mexico and USA; and 5,000 in Belgium and Switzerland.

Other less popular methodologies consist of counting agglomerations of homes; for example, 100 contiguous dwellings constitutes an urban area in Peru. Other countries consider the availability of services as defining urban. For example in Honduras, an area is urban if, in addition to having a population of 2,000 inhabitants, it possesses services of water, electricity, education and health infrastructure. This definition is relevant from a poverty analysis standpoint because the absence of these services is usually associated with poverty; thus, this particular definition of rurality provides elements for characterizing poverty. In comparison, rural areas in developed nations have availability of all of the above mentioned services, spatially more scattered, but still available.

The limitations in comparability imposed by the differing definitions of rurality have spurred efforts to create internationally comparable measures of rurality. These projects merge satellite imagery, which shows population agglomerations, together with census data to spatially distribute populations. The GRUMP project (Global Rural Urban Mapping Project) is one major effort of the type, identifying cities with night-light satellite imagery and using census data to distribute population inside and outside those cities. Although the effort is important, it is still not an ideal measure because it depends on the reliability of the original census (and its degree of sub-administrative unit detail), and because it uses official (and therefore different) rurality figures in the model used to assign rural and urban populations.

CHANGE IN POST-INDEPENDENCE SCENARIO

In the half century or so of independent existence, South Asian states have given varying degrees of attention and priority to rural development. Moreover, the motivations and objectives of public policies have varied over time and among countries owing to a complex combination of political, economic and social factors as well as opportunities for development provided by foreign assistance and global economic trends. Nevertheless, there have been some uniformities in the rural development policies of South Asian countries because of the similarities in their social and cultural milieu and a shared legacy of their colonial past which included institutions of governance and an agrarian structure suited to the needs of the colonial rulers.

The objectives of rural development followed and implemented by the various Governments have also been influenced by the overall plans for development and the performance of the economy in its global setting during this period. In particular, they have been conditioned by the international economic environment, especially for foreign assistance. Although rural development has been a priority area for external donors its importance has increased in the second half of the last 50 years as concerns about food security, the population explosion, environment and climatic changes, as well as poverty, equality and social justice have come to be increasingly perceived as being in the purview of global, rather than national policy agendas. Rural development stood at the cross-cutting path of these concerns. However, the multiplicity of the objectives which rural development was expected to achieve often deprived it of a central focus and often contributed to its failure.

RAISING AGRICULTURAL PRODUCTIVITY

This has been – and to a large extent continues to be – the primary objective of and the principal motivation for most rural development programmes, which were undertaken in the wake of rising population pressure on the land and the need for transferring resources for the pursuit of economic diversification. Before the advent of the green revolution in the 1960s most South Asian countries faced chronic food shortages, in some cases actual famine, and had to import substantial amounts of foodgrains, often financed from PL 480 imports from the United States of America and assistance from other countries.

It needs to be recognized that while raising agricultural productivity is an essential goal, it is a means for achieving larger development goals, such as increased welfare and the alleviation of poverty of the population. While increased agricultural productivity may ensure abundance of food availability for a given population, it may well not provide an adequate nutritional food intake for the majority of the population or, much more important, it may not provide sufficient access to those who need food, despite large surpluses of foodgrains.

A striking example of this paradox is Brazil, which despite being one of the leading exporters (especially of soybeans and orange concentrate) is one of the most ill-fed nations in the world (Souza Silva, 1994). The paradox is no less evident in India, where as a result of the Government's successful drive to raise food production and stocks about 70 million tonnes of wheat and rice lie in Government godowns while over 200 million children, women and men remain chronically undernourished. Pregnant women are the worst affected, since maternal and foetal undernutrition results in the birth of children with low weight (less than 2.5 kg). Such children are handicapped at birth in mental development (Swaminathan, 2002).

Also, as pointed out by Rao (2002) in the context of India, foodgrain security, though essential, cannot be equated with food security. The share of consumer expenditure on cereals now accounts for a little less than 40 per cent of total consumer expenditure on food in the country, the remaining 60 per cent being incurred on items such as edible oils, sugar, milk, eggs, meat, fish, vegetables and fruits. Even for the poor these non-foodgrain items account for as much as half the total expenditure on food but in order to bring the intake of these items to adequate levels their consumption by the poor has to increase at least threefold. The demand for these items of food will therefore rise at a high rate due to population growth as well as the rise in per capita income. The goal of food security, therefore, goes far beyond attaining self-sufficiency in foodgrains and should aim at attaining physical as well as economic access to a balanced food basket, especially for the poor. However, achieving economic access to non-foodgrain items would require a much stronger effort to raise the purchasing power of the poor than ensuring the necessary supplies.

There is an urgent need for diversification of South Asian agriculture from its current focus on foodgrain and cash crop production to the production of non-cereal products, which will also serve to raise employment and increase the purchasing power of the poor, to a considerable extent. This is because the potential for employment generation in dairying, horticulture, etc. is much greater than in cereals. These activities also have the potential for greater human resource development which would in turn result in higher wage rates and provide the necessary purchasing power for both foodgrains and non-foodgrain consumption of the poor.

The objective of raising agricultural productivity in the context of rural development programmes is often postulated without any reference to the agrarian structure prevailing in a country. However, there seems to be a persistent bias in such programmes towards the larger farmers. As pointed out by Banerjee (2001), few historical phenomena share this remarkable tendency in the history of agrarian relations. "The state, it appears, has intervened always and everywhere in the markets for land, agricultural labour and other inputs into and outputs from agriculture to make life easier for larger farmers". Such a bias could be defended as a means of achieving food security if it could be demonstrated that large farmers were in fact more efficient than small farmers. On the contrary, however, there exists a large body of evidence to show that small farms in developing countries, including South Asia, tend to be more productive than larger farms. The logic of the argument about the higher productivity of smaller farms is quite simple and is based on the higher costs of supervision of hired labour in larger farms and the relative scarcity of land in relation to the availability of family labour virtually at zero opportunity cost (Bhaduri 1997). The smaller farms are also able to grow additional crops and engage in subsidiary activities to supplement their incomes and for the survival of their family while the larger farms concentrate on only one major crop.

PROMOTING A SUITABLE ENVIRONMENT FOR THE UPLIFT OF THE RURAL COMMUNITY

Among the most fundamental changes in the evolution of rural development programmes is the continuing debate about the need for a change in their ethos and the way the protagonists of these programmes (who are often outsiders) relate to their beneficiaries. Most of the earlier programmes were either paternalistic in nature or were run by self-serving bureaucrats who were often closely allied to elite groups. There are still very few programmes of rural development which live up to the motto of being of the poor, by the poor or for the poor. The induction of NGOs in poverty alleviation and rural development programmes has to some extent brought them closer to the ideal but the empowerment of the poor is still more rhetoric than reality in the South Asian context. In order to empower rural communities, the rural development programmes along with the Governments and institutions of civil society need to focus on a number of interrelated areas, particularly on human capital development. Most South Asian countries are already committed to goals such as education or health for all, which should remain on every agenda. What is needed is not only the speeding up of the implementation of these goals, but also of moving beyond them in the direction of ensuring some affirmative action in favour of the poor.

The priority should be to move towards substantially enhancing investment for the purpose of upgrading the quality and governance of rural schools and health care facilities to a level where the rural poor do not feel disadvantaged compared to the urban middle class. Such a goal carries formidable implications as to costs and governance and may need some deployment of resources from non-priority projects. However, the resultant effect of such a process of quality enhancement efforts could enable the younger generation of poor households to compete on the basis of equality with the children of the elite for places in the universities and in employment. This would transform the competition between the children from poor and elite households into a more level playing field and would have "an empowering effect" on the poor to demand more rapid democratization of opportunities for human capacity development.

PROVIDING AND HELPING ACCESS TO BASIC FACILITIES AND THE MANAGEMENT OF THE RURAL COMMONS

A common critique of most rural development programmes is that they fail to cater to the needs of the more vulnerable groups such as women, the landless, minorities and other deprived groups of South Asia. To reach and include them requires a deeper understanding of poverty and its underlying causes; an emphasis on building critical human, social and physical assets; and more effective delivery of basic services. For example, despite improvements, access to education and health care remain, along with other social indicators, below desired levels, especially in rural areas and among women and female children. Such poor delivery of basic services works to limit progress in human development.

Besides rural-urban differences in the availability of services, the country averages also disguise variations based on caste and geography. In Nepal, for example, "untouchables" have a life expectancy of 45 years, 15 years less than upper-caste Brahmins. In India adult literacy rates among women of scheduled tribes, most of whom live in rural areas, was 29 per cent compared to 39 per cent for all Indian women. Country averages further mask the wide geographic diversity in these indicators within the country. In nutrition surveys by the National Nutrition Monitoring Bureau (NNMB) of rural areas in India (1994), the percentage of children underweight in nine large Indian states varies from 50 per cent to almost 80 per cent and severely underweight from 15 per cent to 35 per cent. Recognizing the persistence of rural vs. urban, gender, caste or ethnic and geographic biases may help in the design and targeting of development programmes. These imbalances can be reduced by interventions through rural development programmes that focus on increasing the access of the poor and vulnerable groups to essential services that are presently available only to the privileged and elite groups. In order to respond effectively to the needs of these vulnerable sections of the population, rural development programmes must have the following features:

(a) The adoption of decentralized, participatory and beneficiary-driven approaches designed to improve the delivery of such rural services as drinking water and rural sanitation, irrigation, extension, microcredit, education and health to the poorest sections of the population;
(b) Community management for the sustainable use of natural resources, such as joint forest management and watershed management programmes;
(c) Fiscal and administrative decentralization to local governments for enabling them to undertake the programmes identified above;
(d) Measures to improve governance and social inclusiveness of public sector institutions across income, gender and ethnic groups;
(e) Measures to reduce the vulnerability and risks faced by the rural poor and measures to help them recover from natural catastrophes such as floods, droughts and hurricanes (e.g. disaster and coastal management) and to improve the effectiveness of existing government safety nets (e.g. food and rural employment).

These measures will require the combined efforts of the Government (at the national, subnational and local levels) as well as foreign and private donors and non-governmental organizations, including the rural support programmes.

BRIDGING THE RURAL-URBAN GAP

Rural-urban disparities have been an endemic problem of most developing countries and a major source of the continuing increase in their urban population. Although the main reason for the rural-urban population drift is the push of rural unemployment and underemployment, a contributing factor is also the lure of greater access to the amenities of life which is in inverse proportion to the distance from urban metropolitan centres.

Enhancing The Role of Technology as a Means of Raising Productivity

Facing the increasing scarcity of arable land, South Asia must apply technological innovations to improve efficiency and sustain productivity growth. If the region's population nearly doubles, as projected from 1.4 billion in 2000 to 2.2 billion in 2045, the challenge of keeping agricultural growth rates at par with population growth will require putting available technology to better use and, through energetic research, developing new and more efficient growing methods. The average rate of agricultural growth per annum during the period 1990-98 was 1.5 per cent in Sri Lanka, about 2 per cent in Nepal and Bangladesh, 3.8 per cent in India and 4.5 per cent in Pakistan. This pace is particularly worrying as there are indications that the rate of growth of total factor productivity, despite considerable potential, is slowing down in many areas.

In Pakistan, for example, it is estimated that large, existing productivity gaps in major crops indicate an opportunity to boost productivity by as much as 30-40 per cent over the short to medium term. This projection is based on wider diffusion of available, improved crop production technologies, more efficient use of land, water, and other inputs and better post-harvest handling of produce.

Other South Asian Countries can Score Similar Advances

While the benefits of the green revolution are generally recognized to be scale-neutral and did benefit the poor in South Asia to a considerable extent, the gains from technological innovation remain unequally distributed between those with access to land, water and inputs, and those without. There is broad consensus that the main causes of rural poverty lie in low rates of agricultural growth and factor productivity and that the key to raising productivity in agriculture lies largely in measures to broaden access to land and complementary inputs, along with a more favourable policy environment towards agriculture (Fan and others, 1998). More equitable distribution of operational land holdings would create more equitable patterns of demand, which in turn would enhance growth in the rural non-farm sector and remove some of the biases in credit, marketing and research institutions that arise from the unequal distribution of assets and power

(Singh, 1998). This is supported by recent evidence which suggests that countries with a more equal land distribution experience higher rates of economic growth.

The knowledge and information revolution is now being brought within the reach of the remotest areas by advances in telecommunications and information technology and needs to be harnessed not only for the rural elite but also for the rural poor. Formidable opportunities are being opened up in the area of distance learning and medicare, for urban standards of education, medical diagnosis and prescription to be delivered to the most remote villages. Here major investments to build the infrastructure to take the IT revolution to the villages, remains an important goal of public and global development policy. It also provides an opportunity for collaborative arrangements between the public and private sectors, as well as the NGOs. The example of Grameen phone in Bangladesh enabling poor, rural women to be brought into the communications revolution, as both providers as well as users of IT services, needs to be emulated elsewhere in South Asia.

THE RURAL-URBAN DIVIDE

The rural-urban divide is not a discrete attribute but is a continuum, ranging from the urban suburb or periphery to the most isolated or distant rural communities. Rural development programmes need to be fine-tuned to take account of the specificity of the problems that are faced by communities lying between the two ends of the rural-urban spectrum. At least four major divisions of this spread with their own distinctive rural strategies can be distinguished.

Peri-urban Areas

The main issues arising in the urban periphery are not dissimilar to those in urban squatter settlements: creation of jobs in industrial and service sectors in neighbouring urban centres, provision of adequate transport facilities and housing. Many of these areas have been reduced to dormitories of adult residents who commute daily on bicycles or animal-driven vehicles to the urban centre. To the extent there still exists some scope for farming in these areas, the rural development programmes can help promote micro-scale, high-value-added farming, such as vegetables and dairying which would provide fresh produce, create jobs and avoid pollution. There also exists considerable scope for rural industrialization in these areas which could take advantage of backward linkages with agriculture and forward linkages with urban industry. For this there will be need for credit and new credit institutions which could provide venture capital for the establishment of small-scale industries and services to enterprising individuals or groups.

Accessible Rural Areas with Good Natural Resources

These areas are good candidates for agricultural development with the help of market incentives and institutional development. They have potential for higher absorption of both labour and capital and of producing market surpluses. With investment in human development and technology, these areas could become highly productive and their products could compete in world markets. These areas could provide employment to people of other less well-endowed areas, especially during seasonal peaks.

Accessible Rural Areas with Poor Natural Resources

The possibilities of productive employment are likely to be low in these areas and migration may be the only alternative for most people. However, possibilities of livestock farming could be considerable and may provide opportunities for employment in dairying and related activities, along with handicrafts for women. The main handicap is likely to be access to water and capital investment in tubewells and small irrigation projects are likely to yield beneficial results, as land itself is unlikely to be scarce. Since these areas are not remote it may not be difficult to access services from Government and non-Government organizations engaged in rural development activities. In particular, they could receive the services of teachers and health workers for training people and providing basic education.

Remote or Isolated Rural Areas

These are the most difficult areas to deal with as the costs of construction of infrastructure to reduce their remoteness are generally high. Nonetheless, measures to improve their productivity and incomes can be undertaken by subsidizing certain economic activities such as poultry farming and livestock. They could also be assisted in launching public works programmes to help build the needed infrastructure. Some remote areas have the advantage of being yet unspoilt by excessive human habitation and still preserve their pristine beauty. They could become attractive destinations for eco-tourism. Inhabitants of the area could be encouraged to preserve and protect wildlife and biodiversity and to guard against poaching by illegal hunters and fortune seekers.

THE ROLE OF NGOS IN RURAL DEVELOPMENT

The success of the Rural Development depends upon the active participation and willing co-operation of the rural people through self-help organizations and voluntary agencies. In recent years, the voluntary agencies have acquired greater importance and significance than before.

Voluntary action stimulated and promoted by voluntary agencies engaged in development play a significant role at the grass roots level in the Indian social milieu; The NGOs and Government Departments are also struggling hard to organize the people through social mobilization process for eliciting their participation in the successful implementation of the development programmes. Some have succeeded in the process and some are still struggling in progressive direction.

Voluntary organisation is not a new phenomenon in our country. Voluntary effort has always been an important part of our culture and social tradition. The need for organizing people into accredited associations and their involvement and participation in rural development have now been fully recognized. In recent years, they have increased in considerable number, acquired greater importance and significance and put up many new experiments in the field of rural development.

Voluntary organisation can play a crucial role in rural development by supplementing government efforts as they are close to the minds and hearts of the rural people. They have their roots in the people and can respond to the needs and aspirations of the community very effectively. They can experiment new approaches to rural development. (Dhillon & Hansra - 1995).

The success of the rural development depends upon the active participation and willing co-operation of the rural people through Self-

Help organizations and voluntary agencies. In recent years, the voluntary agencies have acquired greater importance and significance than before because the administration has not been able to reach the people, especially the poor and weaker sections. They have been able to make their presence felt from the local to the national level and now at the international level also. Many of them have pioneered works in areas, which were ignored by the process of national development planning (Anandharajakumar - 1995).

Various Role and Functions of NGOs

Voluntary action stimulated and promoted by voluntary agencies engaged in development play a significant role at the grass roots level in the Indian social milieu; the success of rural development depends upon the active participation of the people through Self Help Organizations. The various roles of NGOs are described below for better understanding.

a. Catalyze Rural Population
b. Build Models and Experiment
c. Supplement Government Efforts
d. Organizing Rural Poor
e. Educate the Rural People
f. Provide Training
g. Disseminate Information
h. Mobilize of Resources
i. Promote Rural Leadership
j. Represent the Rural People
k. Act as Innovators
l. Ensure People's Participation
m. Promote Appropriate Technology
n. Activate the Rural Delivery System

The widespread belief that NGOs are more successful in reaching the poor in poverty reduction also resulted in rapid growth of funding for NGOs by government and external donors. As far as the government funding is concerned, there are over 200 government schemes initiated by the central and state governments through which NGOs can have direct access to resources for rural development (Reddy and Rajasekhar 1996).

THE CHANGING FACES OF RURAL LIVELIHOODS IN INDIA

India is witnessing a series of changes since early nineties. Recently, the Sensex crossed 20,000 points and simultaneously India ranked in 94th out of 118 countries in the Global Hunger Index — behind Ethiopia. Unprecedented numbers of farmer suicides, big corporate houses entering into retail business, land allocation for Special Economic Zones, boom in information technology and IT enabled services, zero growth rates in employment, are among many such events that needs introspection.

In recent years the Indian government has made huge investments in development of infrastructure like roads, telecommunication, etc. It has also passed legislation to benefit rural citizens. For example, under universal service obligation, each village should have a village public telephone installed and maintained by the service providers. Also, the "Right To Information Act" passed recently was to make the government system accountable to the citizens. However, civil society organizations are often unaware of such legislation and they fail to leverage the benefits. Corporate houses, private business houses and largely urban citizens are making effective use of the provisions. Thus, the benefits are inequitably distributed between the rural and urban areas. This rural-urban divide in accessing infrastructure services coupled with inability of civil society organizations to utilize the existing provisions has contributed to the slow growth of livelihood opportunities in rural areas.

Initial poverty eradication efforts in India concentrated on supply of agricultural technologies, inputs and services that were often 'production' orientated. However, they were not targeted towards the poor or were largely inappropriate to the needs of the poor and the benefits were mostly captured by the wealthy. Later, the approach changed towards 'capacity-building' in sector organisations to equip people and organisations with the skills and resources to do a better job but, overall little has changed since the new skills are not used. All these approaches tended to be sectoral and supply-driven and the results were not very encouraging. The concept of livelihoods and livelihoods analysis emerged in the mid nineties - closely associated with poverty reduction strategies. Understanding the livelihood systems of the poor is crucial to effective poverty reduction. Livelihoods of the poor can never be understood in any one-track logic - be it economic, social, technical, cultural or political. The livelihood systems are made up of very diverse elements which - taken together - constitute the physical, economic, social and cultural universe wherein the families live (Hogger, 2006). Thus, the livelihood system is more than just a set of physicoeconomic reconditions for continued existence. It also encompasses psychosocial dimensions of experience of living. The livelihoods approach puts households of the poor as its central focus. It takes holistic consideration of things that the poor might be vulnerable to, assets and resources that help them thrive and survive, policies and institutions that impact their livelihoods, how the poor respond to threats and opportunities and what sort of outcomes the poor aspire to.

Agriculture and allied activities support livelihoods of nearly 70 percent of India's rural population. In recent years, land based livelihoods of small and marginal farmers are increasingly becoming unsustainable, since their land has not been able to support the family's food requirements and fodder for their cattle. As a result, rural households are forced to look at alternative means for supplementing their livelihoods.

Changing Trends in Agriculture

Majority of the India farmers derive their livelihood from agriculture. During the Tenth Five Year Plan, gross domestic product (GDP) originating from agriculture and allied activities was 2.3 percent compared to 8.0 percent in the industrial sector and 9.5 percent services sector. During this plan period, the growth in the agriculture and allied activities averaged 2.3 percent which is lower than that of 3.2 percent during the 1990s and 4.4 percent during the 1980s. Also, there is a shift from staples to cash crops which is the major reason for food insecurity. From 1960-61 to 1998-99 the area under grain crops has gone down from 45 million hectares to 29.5 million hectares, area under cotton has increased from 7.6 to 9.3 million hectares and area under sugarcane has increased from 2.4 to 4.1 million hectares. Since 1990-91, due to the new economic policies, the area under food grains and coarse grains have declined by -2 and -18 percent respectively while area under non-food cash crops such as cotton and sugar-cane have increased by 25 and 10 percent respectively. However, production of milk has increased from 84.4 m tonnes (2001-02) to 97.1 m tonnes (2005-06). Production of eggs has increased from 38729 millions (2001-02) to 46231 millions (2005) Ghatak, 2007). Notwithstanding increased availability of milk, fruits, vegetables, fish and other produce, the agricultural sector is facing the new challenges of diminishing land resources, factor productivity decline, threatened loss of bio-diversity, natural resource degradation, widening economic inequality, etc that have serious implications on the livelihoods of the poor.

Indian agriculture has also come under significant adjustment pressure from market liberalisation and globalisation. The changing consumer preferences have added to the complexities of adjustment. The food consumption pattern in India is diversifying towards high value commodities. The decline in per capita consumption of cereals, in particular coarse cereals, has worsened the nutritional status of the rural poor. In the case of the poor, total consumption of high value cereals like rice has declined by 10 percent due to rise in prices of cereals in real terms during the 1990s and dietary diversification towards non-foodgrains. Similarly, average daily intake of protein by the Indian population decreased from 60.2 grams to 57 grams in the rural area between 1993-94 and 2004-05 (Praduman Kumar, et. al. 2007).

Migration and Livelihoods

Every year millions of poor families migrate in search of work. They are forced to migrate due to a livelihoods collapse in the villages. These distress migrants often lock their homes, take a few meager belongings and move across long distances. The children accompanying their parents are forced to drop out of school. The numbers of such children under the age of 14 years is estimated to be around 9 million.

Being away from their village, they do not belong to the places where they go and increasingly lose acceptance in their own villages. They are disconnected from their community, culture and traditions, unable to take part in festivals, fairs, religious and social functions, which are an integral part of their lives thus losing their sense of identity. The vulnerability of people who cross state boundaries is even greater as they find themselves increasingly at the mercy of their contractors. The family members or children left behind in villages do not know where the family has gone, or how to contact them. Dealing with emergencies, particularly back home is difficult, especially for those who go long distance. Distress migrants find themselves with hardly any skills, assets or education. Earnings from migration are survival level and not enough to build assets. On the contrary, medical or other personal emergencies and the unsteady nature of work often push them further into debt. Those who own a few animals have to arrange that they are cared for or just end up selling them off at low prices – thus depleting their assets.

The work sites largely have no basic facilities like water, marketplace, schools, health centres, etc. Migrants are at the mercy of their employers for all their needs. Contractors retain control of labourers not only financially but also physically, forcing them to work even when they are sick or injured. If there resident communities nearby, migrants are usually shunned by local people, and regarded as bad elements or thieves.

Given the difficulties encountered by illiterate and under-informed migrants in accessing provisions that have been made for them by the State, there appears to be a need for support from elsewhere. A number of civil society organisations have taken up the cause of building the capacity of migrants to demand their rights from contractors and government officials.

A particularly striking example of a successful migrant support programme is the one initiated by the Grameen Vikas Trust (GVT). It has worked closely with the panchayats of source villages and has developed an informal system of identity cards for migrants. With these cards, migrants have something to show the authorities at railway stations and bus stands, common points in their journey when they are open to harassment. GVT has liaised with NGOs in the neighbouring state of Rajasthan to set up migrant resource centres that provide them with information on job availability, wage rates and rights. By all accounts such migrant support programmes have achieved more success than official programmes in ensuring fair pay and better working conditions. The widespread system of subcontracting that operates in non-farm sector employment especially in construction projects fudges responsibility and accountability so that it becomes very difficult for government officials to implement protective legislation.

Trends in Rural Non-Farm Sectors

The rural non-farm economy is the backbone of numerous towns scattered throughout India as well as the primary source of income and employment for many of India's poor. The sector therefore plays a key role in determining future prospects for employment growth and poverty alleviation in the country. The RNFS is closely linked to agriculture. A large share of rural manufacturing involves agro-processing and the production and supply of farm inputs. Agricultural performance therefore tends to influence growth in the non-farm economy. A growing agriculture demands production inputs in addition to supplying raw materials to transport, processing and marketing firms. Moreover, agriculture provides the livelihood of more than 70 per cent of the rural labour force and as such, increases in farm income tend to stimulate demand for consumer goods.

Interestingly, development of the RNFS in the 1980s resulted in a remarkable reduction in poverty. However, this is an area which has received casual treatment from the policy makers, economists, politicians, social activists, entrepreneurs and more importantly, the communities themselves. Despite this neglect, it is said that 90 percent of the additional employment during the reform period was due to the non-farm sector. According to the National Sample Survey, the percentage of employment in the RNFS in total rural employment increased from 18.4 per cent in 1983, to 21.6 per cent in 1993-94 and 23.8 per cent in 1999-2000.

The additional employment created in the RNFS during the reform period was 8.54 millions. Out of these, 36 percent of the additional employment in RNFE during the reform period was due to construction, 27 percent was due to manufacturing and repair services and 25.5 percent was due to transport, storage and communication. The contribution of retail trade was 13 percent while that of hotels and restaurants was only 4 percent. In the case of community and personal services, there was a decline in employment while the employment of social services (education and health) increased. The shares show that around 63 percent of the additional employment was due to construction and transport, storage and communications. It is not clear whether the creation of additional employment in these activities led to improvements in the well being of the poor.

Although the RNFS has tremendous potential of removing unemployment in villages, the government has not been serious to address the basic problems plaguing this sector. Major bottlenecks in the holistic development of the RNFS are poor quality of employment and incomes, shortage of skilled manpower, unavailability of credit facilities, absence of marketing networks, poor transportation facilities, low public investment in villages, lack of basic amenities in rural areas, poor law and order scenario, erratic power supply, etc. Besides, not much has been done to develop technologies relevant for the sector. Entrepreneurs are by and large reluctant to invest in the rural non-farm sector. Not many NGOs and political groups are engaged in empowering the rural communities to be proactive and develop RNFS enterprises on their own.

Globalisation and Vulnerability

Rural livelihoods are affected by the process of Globalisation. Special Economic Zones (SEZ) are created across India to promotion exports. Commerce and Industries Minister Kamal Nath claims that exports will ultimately grow five times, GDP will increase by 2 percent , and 30 lakh jobs will be generated by SEZs across India. The government also claims that SEZs will attract global manufacturing through Foreign Direct Investment (FDI), enable the transfer of modern technology and create incentives for infrastructure.

To date, the central government has approves 237 SEZs in 19 states (occupying 86,107 hectares) out of which 63 of these SEZs have already been notified. Around 23 SEZs are operational, 18 in the IT sector. Total amount of land to be acquired across India is estimated to be around 150,000 hectares. This land – predominantly agricultural and typically multi-cropped - is capable of producing close to 1 m tonnes of foodgrain. If SEZs are seen to be successful in the future and more cultivated land is acquired, they will they will endanger the country's food security.

Reliance Industries has sought permission to build India's biggest Special Economic Zone near Mumbai comprising more than 14,000 hectares of land. Similarly, Some 400 SEZ projects, offering incentives to big businesses had been approved (formally and in-principle) amounting to 1,25,000 hectares. Some estimates show that close to 114,000 farming households and an additional 82,000 farm worker families that are dependent on these farms for their livelihoods will be displaced. The total loss of income to farming and farm worker families is at least Rs 212 crore a year not include other income lost due to the demise of local rural economies. Such developments by the MNCs and Corporate Sector are playing with the emotional, economical and cultural interests of farmers, fisherfolk, salt-pan workers and agriculturalists in the area. Their entire livelihood is dependent on the farmland. It is an amazing feature of this exclusive area of India where people have farmlands and they also do inland fishing. India's Special Economic Zones have drawn sharp criticism from a wide range of environmental and human rights groups, as well as farmers, villagers, fishing and agricultural workers who face losing their land, homes and livelihoods. Critics point to a lack of transparency and unwillingness to hold open public consultations on the SEZ issue.

Correspondingly, the entry of corporate retail into India is nothing but a hijack of India's vibrant, well organized retail service economy. Corporate retail will not create two million jobs; but it will destroy 38 million livelihoods of people involved in running small shops and street markets. The growth being projected as a new contribution to the economy hides the destruction of the contribution of the 40 million people involved in small retail to the Indian economy. Corporate retail is being described as organized retail. And small scale decentralized and highly networked indigenous retail is being falsely referred to as unorganized. However Indian trade is highly organized and has existed for centuries on the basis of low cost and high efficiency. India needs the self organized skills of our traders, shopkeeper, hawkers and vendors both to provide employment to millions and affordable friendly community service for basic needs to society. India is a land of retail democracy. Hundreds of thousands of weekly haats and bazaars are located across the length and breadth of our country by people's own self-organizational capacities. India's streets are bazaars – lively, vibrant, safe and the source of livelihood for millions. India has the highest shop density in the world, with 11 outlets per 1000 people not including the village haats.

Our retail democracy is characterized by high levels of livelihoods in retail with nearly 40 million employed accounting for 8 percent of the employment and 4 percent of the entire population. It is predominantly self organized with low capital input and highly decentralised. In a country with large population and high levels of poverty, this model of retail democracy is the most appropriate in terms of ecological sustainability and economic viability. By 2011, more than 6600 mega stores are planned with investment of Rs. 40,000 crore. Reliance plans to invest $5 billion over next four years to open thousands of retail stores. Walmart's partner Bharti also plans to invest up to $2.5 billion in new stores in the next eight years. The entry of the giant corporate retail in India's food market will have direct impact on India's 650 million farmers and 40 million people employed in tiny retail.

ROLE OF EXTENSION IN RURAL DEVELOPMENT

Since the history of mankind, agriculture evolved round the basic food needs of man and developed principles and practices in crop production including field management. With changing situation, technological development have shifted the traditional agriculture from more crop production to mixed farming involving allied activities like animal husbandry, sericulture, pisciculture etc. Need for technological development in agriculture in our country arose due to the fact that -

- The structure of economy remained predominantly agriculture;
- The cultivable land being limited, productivity has to play an important role to meet the domestic and export requirements of farm produce, and
- The development plans also aimed at self-sufficiency in farm production, in particular, foodgrains to meet the growing demand of population.

The technological development in agriculture, necessarily mean the use of modern inputs at micro level to improve productivity of land and accruing desired economic benefits therefrom. In this context, the most important inputs that play crucial role in agricultural development are :-

a. Use of high-yielding varieties,
b. Use of chemical fertilisers,
c. Use of pesticides and insecticides
d. Use of water as input in agriculture-irrigation
e. Use of pump sets, tractors, powertiller etc.

f. Use of mixed farming technique like animal husbandry, pisciculture etc.

Present agriculture development is not product specific but operates within a specific socio-economic system and inter-sectoral dependency. It is, therefore, essential to create impact of the technology in agriculture on the entire farming community to achieve the desired goals of development. The review of agricultural development in the country indicates that 50 million tonnes of foodgrains production capacity just 30 years back has reached around 210 million tones in 2005-06. Indeed it is an impressive achievement, yet we have to reach more than the current. Our primary task is therefore to identify the various known inputs in agricultural and make them more popular amongst the farming community through selective measures such as multiple cropping, use of high-yielding seeds and such other inputs including irrigation. But some of the short comings responsible for slow progress in agricultural development are :-

- Still traditional low yielding varieties are used in dry farming,
- Lack of proper agronomic practices in water conservation, sowing methods, seed rate, fertiliser use and absence of control of pests and diseases.

The benefits of development in agricultural sector have been derived mostly by the affluent community of rural sector, whereas little access to such technologies was available to the weaker sections of the farming community such as small and marginal farmers.

Place of Extension in Rural Development

In this background, it is of utmost importance to play an active role by various extension agencies operating in the rural sector for development of all sections of the population through improved technology in agriculture and allied activities. The extension process is concerned with communicating the technology of scientific agriculture to the farmers in order to transform traditional level of agriculture to better for improving their economic conditions. Hence, extension in agriculture would mean stretching out the knowledge to farmers on adoption of the new technology and improved practices in various sub-sectors like crop production, livestock rearing fodder production, sericulture, bee-keeping, horticulture etc.

Extension aims at changing the outlook and attitude ofthe farming community in general and it seeks means to improve the farm operations and farmer's family life in totality on their own initiative. As the farmers are mostly small and marginal, they lack direct access to developing agricultural technology. Educating such a group of farmers has to be, therefore, a sustained process to keep pace with rapidly changing agricultural technology.

Concept and Objective of Extension

The modern concept of extension process is working with rural people through informal education for achieving total community development covering several activities, agriculture being the most important. Agricultural extension is considered to be a special branch of rural extension dealing with several economic and social aspects of farming community such as:-

- Efficient farm production
- Efficient marketing, distribution and utilisation of farm produce
- Conservation, development and use of natural resources for farm sector development
- Efficient farm management
- Improving the standard of living of rural population
- Development of leadership for community improvement.

However, the objectives of the extension programmes depend upon socio-economic conditions of the people in the area and need to transmit the latest developments in the field of agriculture to suit their requirements. Thus, the objectives have been set out in all such programmes primarily:

- to assist people to identify their problems and felt and unfelt needs;
- to develop leadership among various sectors of the rural community;
- to transmit information based on agricultural research and applied experience with a view to bring about larger farm community under the modern technology;
- to keep research workers constantly informed about the problems at farm level to offer solutions based on further research in the specific area concerned.

Extension is a continuous process of transmitting the knowledge on latest development in agricultural technology to a section of community for field application and feedback process to know the problems arising for further improvement of technology.

Philosophy of Extension

The philosophy of extension is primarily based on (1) development of the individual farmer and (2) interaction within the various sectors of rural community to avoid imbalanced development. In this process the individual farmer is supreme and the first training group that one can think of is the farm family. Secondly, extension services are rendered with a view to plan, execute and evaluate the technological changes adopted by farmers. Thirdly, extension aims at teaching the farming community and motivating them to bring the latest development into practice through a spontaneous response rather than forcing them to take it up. Naturally, certain techniques and tools are required to teach the people how to help themselves and to ensure that extension agency acts as a catalyst or change agent.

Function of Extension

While trying to introduce a new concept or a new idea to the farming community, one has naturally to consider how this would be adopted by a large majority to reap maximum benefit from out of the new technology. Yet, this function cannot be adequately achieved unless we bring desirable change in the human behaviour particularly the farming community in the rural setting. The change in behaviour of the rural community it possible through many ways but the most important ones are :-

(i) Change in knowledge: The latest knowledge and information about the technological development to the farming community help in creating awareness of the facts and thereby likely change in attitudes.

(ii) Change in attitude: The change in attitude is generally difficult in the farming community due to environmental situation in which they operate and also due to certain extent of superstitions towards adopting new technology in agriculture.

Motivation will help change attitude to some extent, but attitudinal change is possible in the farming community under the following circumstances.

(a) Change in age level: Responsibility awareness factor
(b) Change in status: Financial and social awareness factor
(c) Personal and natural crises' Family death, illness, war and such other awareness factors.

As mentioned earlier, group reaction is the most important value in agricultural extension in changing the attitude of the farming community. Most farmers confirm to the attitude of the members of their respective groups, based on social or economic structure in rural society. Relatively few maintain an independent attitudes towards adoption process. The change in attitudes of groups is largely observed;

(a) by sharing common knowledge through radio, television, newspapers and experience of people in other villages of areas.
(b) by sharing in planning process or functioning in a group activity like cooperative society, clubs etc.
(c) By sharing in decision making through group discussions for common benefit of the rural community such as lift irrigation scheme etc.

(iii) Change in skills: Change in skills could be achieved by giving adequate knowledge about the new technology and its likely chances of adoption by various farming groups. Since the skills are either mental or manual, both of them accelerate adoption process in farming community for desired level.

MOTIVATION PROCESS IN RURAL PEOPLE

Motivation is basically an inner drive or impulses or intention to do certain acts in a specific manner to achieve the desired goal. In this sense, the function of motivating rural people primarily depends upon satisfaction of some felt and unfelt needs such as :

i. Desire for security: in terms of economic, social, psychological and spiritual one.
ii. Desire for affection or response: in terms of companionship, social mindedness, confidence, etc.
iii. Desire for new experience - in terms of adventure, new ways of doing the existing practices.

Therefore, it is worthwhile to analyse the felt and unfelt needs of the farming community in rural sector which accelerate motivation process through satisfaction of such needs. While introducing a new concepts or a new technology to the farmers, the most important consideration is how to ensure satisfaction of the felt needs of the larger section of farming community for putting it into action. However, individual farmer is motivated through satisfaction of a specific need within the need hierarchy which is in the order of.

iv. Basic need: psychological, food shelter clothing.
v. Security need: economic or social.
vi. Love and belonging need: companionship.
vii. Esteem need: status in the society and within groups etc.
viii. Self actualisation need: Self confidence, self-satisfaction with some spiritual element.

Every individual is placed at one of the hierarchy levels and for motivating him one has to identify the specific need satisfaction of the farming community for favourable acceptance of the new technology to be introduced. Yet, an individual ceases to get motivated on satisfaction of one need and not desirous of fulfilling the next need in the hierarchy.

Another way of motivating different farmers' groups could be through proper planning of adoption process within the entire community. Some of the essential steps of planning adoption process are :-

i. Attention: Farmers should get direct attention to new and better farming techniques through use of various extension tools.
ii. Interest: To sustain interest in the farming community. One idea at a time should be introduced to avoid lack of interest.
iii. Desire: The farming community adopts new farm technology only when there is significant urge for satisfying the specific need.
iv. Conviction: The farming community should acquire confidence in the new technology by doing themselves and creating ability within to do things.
v. Action: The conviction of the farming community is generally converted ultimately into action process in adoption of new technology.
vi. Satisfaction: The final product of the earlier steps is satisfaction that results in adopting the new technology.

RURAL DEVELOPMENT POLICY 2007-2013

With over 56 % of the population in the 27 Member States of the European Union (EU) living in rural areas, which cover 91 % of the territory, rural development is a vitally important policy area. Farming and forestry remain crucial for land use and the management of natural resources in the EU's rural areas, and as a platform for economic diversification in rural communities. The strengthening of EU rural development policy is, therefore, an overall EU priority.

Why Have a Rural Development Policy?

The European Union has an active rural development policy because this helps us to achieve valuable goals for our countryside and for the people who live and work there.

The EU's rural areas are a vital part of its physical make-up and its identity. According to a standard definition, more than 91 % of the territory of the EU is "rural", and this area is home to more than 56 % of the EU's population. Furthermore, the EU's fantastic range of striking and beautiful landscapes are among the things that give it its character – from mountains to steppe, from great forests to rolling fields.

Many of our rural areas face significant challenges. Some of our farming and forestry businesses still need to build their competitiveness. More generally, average income per head is lower in rural regions than in our towns and cities, while the skills base is narrower and the service sector is less developed. Also, caring for the rural environment often carries a financial cost.

On the other hand, the European countryside has a great deal to offer. It gives us essential raw materials. Its value as a place of beauty, rest and recreation – when we look after it – is self-evident. It acts as our lungs, and is therefore a battleground for the fight against climate change. And many people are attracted by the idea of living and/or working there, provided that they have access to adequate services and infrastructure.

This means that the EU's Lisbon Strategy for jobs and growth, and its Gőteborg Strategy for sustainable development, are just as relevant to our countryside as to our towns and cities.

The EU's rural development policy is all about meeting the challenges faced by our rural areas, and unlocking their potential.

Why Have a Common Rural Development Policy?

Theoretically, individual EU Member States could decide and operate completely independent rural development policies. However, this approach would work poorly in practice. Not all countries in the EU would be able to afford the policy which they needed. Moreover, many of the issues addressed through rural development policy do not divide up neatly at national or regional boundaries, but affect people further afield (for example, pollution crosses borders all too easily; and more generally, environmental sustainability has become a European and international concern). Also, rural development policy has links to a number of other policies set at EU level.

Therefore, the EU has a common rural development policy, which nonetheless places considerable control in the hands of individual Member States and regions.

The policy is funded partly from the central EU budget and partly from individual Member States' national or regional budgets.

Structure of Rural Development Policy

The essential rules governing rural development policy for the period 2007 to 2013, as well as the policy measures available to Member States and regions, are set out in Council Regulation (EC) No. 1698/2005.

Under this Regulation, rural development policy for 2007 to 2013 is focused on three themes (known as "thematic axes"). These are:

- improving the competitiveness of the agricultural and forestry sector;
- improving the environment and the countryside;
- improving the quality of life in rural areas and encouraging diversification of the rural economy.

To help ensure a balanced approach to policy, Member States and regions are obliged to spread their rural development funding between all three of these thematic axes.

A further requirement is that some of the funding must support projects based on experience with the Leader Community Initiatives. The "Leader approach" to rural development involves highly individual projects designed and executed by local partnerships to address specific local problems.

As before 2007, every Member State (or region, in cases where powers are delegated to regional level) must set out a rural development programme, which specifies what funding will be spent on which measures in the period 2007 to 2013.

A new feature for 2007 to 2013 is a greater emphasis on coherent strategy for rural development across the EU as a whole. This is being achieved through the use of National Strategy Plans which must be based on EU Strategic Guidelines.

This approach should help to:

- identify the areas where the use of EU support for rural development adds the most value at EU level;
- make the link with the main EU priorities (for example, those set out under the Lisbon and Göteborg agendas);
- ensure consistency with other EU policies, in particular those for economic cohesion and the environment;
- assist the implementation of the new market-oriented CAP and the necessary restructuring it will entail in the old and new Member States.

CHARACTERISTICS OF RURAL SETTLEMENT

- The characteristics of rural settlement are identical to the nature of the rural landscape. Rural settlement is a constant process of adaptation between settlers and the land, resources and types of vegetation and topography. These forms of adaptation give rise to a distinctive rural culture. The main way of organizing settlement characteristics in a rural environment is to go from the most general to the more specific.

Land Use

- The most obvious and important part of rural settlement is the use of the land. This is dictated by the nature of the topography and the available resources such as good soil and access to water. This is the primary form of adaptation—the connection between rural resources and the use of the land. All others derive from this.

Spatial Organization

- The spatial organization of the settlement reflects the existence or non-existence of resources for farming, hunting or fishing. These are manifest in the forms of field use and the network of roads that derive from the shape and mode of the fields. The forms of land use are based on the availability of resources. Slash and burn exists almost exclusively where land is plentiful, and two- and three-field rotations are more characteristic of rural settlement and reflect the advances of agricultural technology.

Culture

- Rural culture derives from these adaptations. One of the most common is the harvest festival, which reflects the communal organization of harvesting, the most difficult and labor-intensive part of rural life. In general, rural culture celebrates a tight community that exists to serve one another in the interests of individual and cultural survival.

Consequences

- All three of the above characteristics of rural settlement take the forms of specific systems of transportation, the boundaries that come to exist among property owners and the clusters of buildings, cultural norms and land use that come to act as the ingredients of the rural settlement and its clearest characteristics.

✱✱✱

5

Media, Technology and Rural Development

INTRODUCTION

Eighty per cent of India, problems are in rural areas. This is at the rook of exhorting the industry, banks, professionals and intellectuals to help accelerate rural development. However, the role of press, in rural development is seldom mentioned. Nor is Press evincing much interest in the problems of rural India.

Except devoting some spare to a agricultural statistics, the mediate by and large, silent on such issues as rural credit. rural banking and, Insurance, rural roads, rural transport and communication, rural industrialisation and electrification, rural water, rural health, rural housing, rural marketing and various rural schemes it is doing precious little to evoke national interest, much less rouse national consciousness on the plight of the rural community. If nettling else, the fact that development strategy today is tilted in favour of rural development should have prompted the press to pay more attention to the problems facing rural India. But It is not so This can be attributed mainly to its culture. Press in India is seedy divided into English press and Language press. In the Indian context, such a division is inevitable, of course.

English pleas is big-city-based and serves largely the needs and tastes of a Class of people who are too busy to he concerned about conditions in Indian villages. The more sophisticated among its readers display an allergy even to discuss the subject.

Mindful of what the preference and expectation of Its readers are, English press has no apace to spare to discuss happenings in rural areas, except major natural calamities, sensational crimes and mass-scale unrest. The readership of the language press Is limited to a particular State or region. By this very fact, it cannot make much impact on the national scene, mobilise national opinion, Further, it is weak In resource which places limitations on its news coverage and reporting. There is also the fact that a newspaper can survive only If it serves the immediate Interests of its readers and advertisers.

RURAL DEVELOPMENT

The plight of rural masses deserves national understanding and Press Is the best agency to bring them about. There are no two opinions today that rural population has so far received a raw deal and to the vast majority, life is an ordeal. Happily, there are efforts to correct past lapses and imbalances. The question is, how can the press be a partner in or at least provide a helping hand to efforts. First of all, the press has to accept the view that it has ought to play a role in rural development.

To begin with, it has a climate that prompts and inspires everyone to pay more attention to rural development towards the press ought to be informer and motivation. Is has to inform end If necessary to educate particularly planners, policy makers, professionals sod Intellectuals on the potential.

Experience of industrial development its India has demonstrated that though high technology is desirable and indeed necessary in certain areas, the war against poverty can be fought only by fully utilising the great asset of India - the abundant manpower. Of course, where the introduction of improved techniques helps increase local employment and raise the level of real earnings of the agricultural labourers and artisans, they are to be adopted. It is in this context that promotion of village and small scale industries forms one of the major planks of the new industrial policy. But availability of money and tools alone will not ensure the success of the programmes for rural development. It calls for a new attitude and awareness of new possibilities among rural peoples It also calls for Inculcating a scientific temper among them. Science involves among other things, an attitude which helps people to get rid of ignorance, superstition and aversion to changes. In this context, India's rural community needs science. Press on which people depend heavily for information can play a purposeful role in making people aware of the changes and new pos sibilities so that rural people do not statistically accept their poverty and backwardness.

Today the press Is doing precious little to educate people on this score, May be there are constraints. But basically, it is a Question of taking a wider view of Its obligations towards society and Its involvement in the tasks connected with rural development. Till now it was thought that Community Development Programme Panchayat Raj and Small and Marginal Farmers' Development Agencies and the like ran usher in new economic order in rural India. Few thought banks financial institutions and research organisations had any role to play In rural transformation. The latter concern on the part of the press the failure of a reform considered to be India's silent agrarian revolution was conspicuous by its absence.

Rural Development is need of the hour for a country like India, which is on the threshold of being a big world power with the capacity to influencing the course of events on our planet. However, this is a dream which we can realize only when we are a country with no internal dissensions and strife, a country with affair economic system and cohesive social order .In fact the essence of development lies not in a regimented system where none dares to disagree but in all -inclusive dispensation in which every one plays an equal role. The spread of education and general awareness on one hand and the revolutionary changes in communication and information technologies on other hand have made media the most potent means of mass education and mass mobilistaion in our time. Media is now being increasingly viewed as the most effective and sought after means of communication. In fact media now influences all works of our life -personal, social and professional. Rural development is need of the hour for a country like India, which is on the threshold of being a big world power with the capacity to influencing the course of events on our planet. However, this is a dream which we can realize only when we are a country with no internal dissensions and strife, a country with a fair economic system and cohesive social order. In fact the essence of development lies not in a regimented system where none dares to disagree but in an all -inclusive dispensation in which every one plays an equal role. The quest for rural development must lead us in the direction of empowering those section of society who are in the need of being empowered. This is the real meaning of democracy and this is the area in which media can and media must play an important role.

To understand what role media can play in rural development we have to understand media scenario in all its dimensions, We have to understand scope, reach and impact of media on society as well as the functions it is supposed to perform. We also have to understand how media influences social change in a developing society. But before this it will be worth to see how information technology has changed the face of media as well as the society which media caters to.

The speed of technological developments in the fields of communication has no doubt changed the face of media and society in the past two to three decades. These developments are so fast that the social scientists and researchers are finding it difficult to asses, assimilate understand and interpret their impact on different communities and social groups. The diffusion of new information technologies are making such in roads into the Indian societies that their multiple applications are giving birth to more and more cynacism. Media is a helpless victim of this phenomenon.

After the conversion of the whole world into a global village the new information technologies which have brought the vedid, the sound, the graphics and text together at a great speed and skills, are basically aiming at creating an information society which, in the very near future, will widen the gap between the information rich and the information poor. The unequal pace of technological development between the urban and rural areas, where are both producers and beneficiaries of technology, has increased inequalities between urban centers and rural periphery.

Today 's technology, itself class oriented, tending to increase divisions between social groups. The gap between information rich and information poor has widened. The Indian media is there fore, living in a situation where it does not know where to go. To survive it has to follow the lead being given by the IT revolution. To live up to its tradition it has to show commitment for those sections of society who are becoming information starved.

For centuries our society has been depending upon the traditional means of communication processes and word of mouth. Human relationships, interpersonal communication processes and word of mouth. Human relationships in closed Indian society used to develop in a natural, healthy and productive social environment. But in the early nineties the satellite and cable tevisions created new type of society (so far, mainly in the urban areas.) by bombarding our households with alien messages. The new information technology is creating a pseudo -environment, in which users tend to believe that by acquiring it, they have acquired everything, including wisdom, thought and logic. Most of the receivers fail to understand that what they do get are information packages, consciously or unconsciously designed for creating an atmosphere of artificial intelligence. The receivers also develop a tendency to absorb the information, which is easy to understand. "In a way the television is taking conversation out of the family, the computers are responsible for taking human relation ships away from the society. A society which has been bound together by inter-personal and group communication is sure to feel uncomfortable with the new information culture. However, very few are able to view this as a problem in the process of our development. As we know our media is now entirely information technology driven. As we can see information information technology has a very dynamic relationship with the society and culture, .We see the scenario for media in the world dominated by the influence of information technology. No doubt this technology brings revolution in the modes of communication by making it more diverse., intra and and interactive and mentally more provocative. But the fear is the diversity would lead to regimentation and segmentation, which may further lead to stagnation, loneliness disintegration, insecurity and isolation. All these major psychological issues in relation to development. That is why it can pose a threat to the individual and group identities and local cultures in case of a plural society like ours.

Yet despite all the changes brought about by information on social scene as well as in media remains a dominant factor with its capacity to appeal to the minds of people. Yet we often witness a tendency to take a very narrow view of media and limit our vision by our media use and our media habits. The whole range, scope and use of media has to be appreciated if want its application in solving the problems our society faces from time to time. The whole range and dimension of media has to be understood if we want to know how media can play a role in promoting the culture of development.

In the wake of increasing commercialization, media's role in expressing public thoughts is shrinking and more constraints and barriers are developing in the way of free and fair interaction between the media and the public. Media and the information systems in the developing countries are being governed by dominant concepts of the West. The developing countries can meet the challenge only through developing alternative concepts of media programming, which have roots in their own social and cultural settings. This demands a great amount of will to experiment on the part of entire gamut of media industry.

MEDIA SCENARIO

Media Scenario basically comprises TV, reach radio, reach newspapers, advertising, other print media forms, attempts at social marketing and net newspapers or cyber journalism. Various forms of traditional media which are integrated in to one social system. These forms are not merely entertainment based. They are essential means of education on various aspects of life. They also form part of larger media scenario. Yet independent news media -print, electronic or even cyber remains the most important media for its capacity to scrutinize public affairs without fear or favour and hold us accountable to our jobs and responsibilities..Its good health and right attitude is therefore, very important to all of us In the all important task of national integration the importance of independent news media cannot be over emphasised.

We can understand the media reach in indian scenario by having an idea of the total number of newspaper, journal s, and periodicals as well as the number of radio stations, TV channels, satellite links and media related websites. In the recent years there has been a steady trend towards what we can describe as micro media — the media of small newspaper, limited circulation journals and community radio.

Basic Formulations

- It is difficult to delineate the total range and impact of this vast organized and at times unorganized media scenario.
- No section of India population is in fact totally media starved.

- Going by the modern idiom of information have and information have -nots. The only thing you can perhaps question is the type and quality of information the vast sections of India's population get.
- Indian press, which was in fore front of india's struggle for freedom from the British and has evolved with the revolution of India consciousness and entry of India in the modern world, therefore, still cares more of tradition, historically and content.

India's national radio national radio network All India Radio has 100 regional stations. There are over 200 local broadcasting radio stations, all government owned (AIR), throughout India. Though a lot many private channels have come in the field of TV the national network Doordarshan is still the biggest telecaster with its three regular channels with terrestrial rights.

Let us try to see what sort of a media community all these media forms create in our country.

- India has 70 million homes with TV, of which 25 milion homes have cable & satelite.
- Perpetration
- The installed base of PCs in Indian home however around 3-4 milion
- Cable TV subscribers -30.08 million in 2000 increased by about 79% to 53.88 million by 2005
- According to a study.
- According to an Indian readership survey, the press reaches about 33% of India's population, while TV reaches 47%.
- India is expected to have 25m Internet users by the end of year 2005,up from three million in 2000 and 5.5 million at 2001 beginning.

According to RNI -(press highlights in 2000)

- UP produced the largest no. (8,750) followed by delhi (6,543) and Maharashtra(5,610).
- The largest number of periodicals and newspaper -are brought out in Hindi (19,685) followed English (7,175).
- The total number of registered newspaper has risen from 49,145 as on Dec 31,2000 to 51,900 as on December 31,2004.

But to known how best media can help promote the climate of development, especially in rural areas we must know how our changing society is influenced by media. As we know the change is brought about more through a process of internalization of certain ideas than by any modification in the external environment. In simple words, one's idea is perceived and corrected by learning through interaction or any other process are more responsible for change in individual and group behaviour than concrete structures that may be built around an individual or group. The media seeks to consolidate the process of internalization.

Since media has the power to expose an individual or group of individuals to new ideas and international, its potential as a change agent is perhaps the strongest. Since mass media has a overhelming reach, its potential as a change agent is perceived to have a multiple effect.

Social change is a by -product of man's effort to control his external conditions. In modifying his environment man sets up a double process or social change. Certain social. relationships are imposed on him by his civilistaion, others are imposed on him by his civilization, others are imposed on him by his civilization.

There is an intimate connection between our behalf and our institutions, our valuations and our social relationships. Certainly all cultural change involves social change because the two are closely interwoven. Most of the Developing countries are emerging from a texdal and colonial background. The told socio economic structures are giving way to new ones. The media, which is the mirror of society, has to relate itself to these changes. The media, has to guide the course of action in the developing societies, apply correctives where needed and also mould itself to the changing realities of the time.

The only worthwhile sensationalism in news in the sensation of social and economic change. Unfortunately Indian media has been specializing in the cult of personality. What some person does, how he behaves or what he says is important, not the issues that matter, the causes that move men, the deseparate needs and urges of the people at large, the living condition s of people in villages and tribal areas. The romance of a dacoit's surrender is presented in colourful language, but the deeper roots of dacoities in Central and Northern India are nobody's concern. There are told attempts at unearthing corruption but no investigations into in to the basic causes of corruption, which is a social cancer as widespread as society and is corroding it. The main cause is the desparate search for funds by political parties, which have to fight costly elections in immense electorates, in which the ruling party leads, This party has made no attempt to reduce political corruption which is the source of all other corruption apart from the economic causes including capitalism under the patronage of multinational funds. A media person is a historian in a hurry, but even in his hurry he should not, if he is to be true to his profession, give up the historical sense, forget the historical perspective. He has to combine many disciplines and in recording and interpreting what is happening to the society around him, he has to be a sociologist also..

Media can perform various functions in promoting consciousness for development. Broadly Media can help contribute in three spheres which will in turns make our society cohesive and better organized. The three areas are.

(a) Economic (b) Social (c) Political

We are one of the few countries in the developing world where right to information has been granted to people. Media has been in the forefront of the campaign for the right to information. The right to information gives a real teeth to democracy which will provide a sense of security and power to the beleaguered of society and give them confidence in system. This confidence is essential for national reconstruction.

The media has a symbiotic relationship with democracy, which in rural terms means empowering the weak. The two cannot exist without each other. The essentials of democracy are well informed citizenary, an equal and effective right to every member of society to participate in the affairs of the state and accountability of all those who are excercising public power and utilizing public resources.

Media is the fourth estate of a democratic polity and scrutiny of public affairs and ascertaining accountability of those who are in the power is its job. Beside the media also provides intelligent in puts by initiating public debates and this improves the quality of governmental decisions. A real and participatory democracy in which decision - making is done in full public view is the best guarantee. of achieving all round development.

FUNCTIONS OF MEDIA

Broadly the functions of media can be divided in to three:-

1. Economic

2. Sociological and
3. Political

In the sphere of economic inputs media can help a person find alternative ways of making a living, which would reduces the pressure on land and raise a family's economic status. Media can help create demand for goods and motivate local initiatives to meet rising demands. It can also help sensitise people to broaden the entre preneurial base media's corruption to overall economic development has been fully realized and it has become a self perpetuating process.

In the sociological sphere of activity media aids in the process of status change form heredity to achievement. To do this it has successfully motivated illitrate to become literate. In our feudal society with its sharp caste and class differences media can help shift influence from age and traditional status to knowledge and ability. To the traditional leaders to compete for status retention media motivates to acquire knowledge and adapt to changes. Media can also help bring about greater equality and a greater respect for human dignity and make cultural and soacial change a self perpetuating process.

In the political sphere of its activity media helps in the process of power change from heredity to achievement. It motivates traditional leaders to defend their power by raising their information level and helps the masses recognize their own importance in the power structure and act as a stimulus to political participation.

Media helps the government learn of the needs of the public and plan its programmes and makes the public know of government plans and programes. Media also helps a community achieve power through mutual cooperation, something which is our concern at the moment and makes a political growth a self-perpetuating process..To sum up we can say that media bring about greater equality and respect for human diginity in the political arena.

CULTURE, MEDIA AND RURAL DEVELOPMENT

We can develop when there is unity in diversity in our society there are lot of many diverse cultural streams. However all of them have to be treated with care and respect. However, the biggest impediment in showing respect to diverse cultural streams is our largely misunderstood stress on main stream. Main stream has become a mind set for most of media persons because subconsciously media shared the misplaced apprehensions of the Establishment that diversity is a danger to the security of the country while the truth is on otherside. In the past few years there has been a lot of discussions on culture as an integrating force. The discussion has left a lot of people confused. Media too has failed to grasp the essence of indian cultural diversity with the result that at many times media reporting to cause differences and promote disharmony in society. It is therefore, necessary that before determining the role of media in national integration we should clearly spell out for media what we as a society understand by Indian culture. In this regard the following questions has to be answered:

- What is the real face of Indian culture in its totality? How to blend the old and new in our living culture so that we neither become orthodox while talking about our past values or become rootless while chasing the mirage of modernism
- What is the relationship between mass communication and Indian culture and civilization? What was the form of mass communication in the pre industrial world? How was it directed and what was its value system.
- What is the role and responsibility of means of mass communication like newspaper T.V. radio and advertising in Today India. ? Is the media discharging its responsibility as it should.
- Society and media has a symbolic relationship. What is the real and active forever of this relationship?
- What relationship you have with means of mass communicaton. Is it a one sided relationship in which you are mere a recipient? If not then what should you do to make it a two -way relationship, and how?

With technology increasing defining our lives and with mobility of people, goods and services becoming easier, traditional cultures find themselves in a situation, where their environment is eroded and their relevance is questioned. This environment naturally breeds a sense of insecurity and alienation. The important questions is how can media contribute to preserving these cultures for the sake of promoting development by functioning in this environment.

To capture and express these diversity of our cultural heritage we need a much bigger media base than what we have at present. This will be possible not by the efforts of big media houses owning chain newspapers and venturing in to small towns. It can be possible by taking mass media to masses and involving members of these cultural groups in media production. A lot many such efforts are under way either by individual initiative or by the initiative of genuine NGO workers.

However, there is need to promote these efforts and for this we may need a re-look at our media policy. There is also a need to revise the government advertising policy to support these small initiatives. Moreover, there is need to organize workshops and training programmes to improve the skills of those who are producing small newspapers. But more than any policy initiative that we take it is necessary that we have a mindset that considers small is beautiful, it is, essential. It is necessary that our obsession with big and few must go.

From a PR executive to a bureaucrat to a political leader ask anyone about the press and only three -four big names come up. Media graduates prefer to hang around big metropolitan dailies rather that accept jobs in medium and small papers, where they can get better status and greater work opportunities. While talented journalists rot and frustrate without work in big dailies the middle and small newspapers are manned by pseudo professionals and non professionals. Respective wage boards have institutionalized this anamoly. For those who think that major English dailies are all about our print media it should be rewarding to know that only four English dailies -The Times of India, The Hindustan Times, Indian Express and the Hindu figure in the list of 20 top dailies of country. Moreover, the big newspaper in English, Hindi and a few in regional languages comprise only 32.7% of the total circulation of all dailies.

The resources -starved small press comprises 33% of total circulation of all dailies. The small dailies no doubt have their problems at times they do not adhere to the high norms of professionalism -yet none can deny that collectively they are fore more representatives of our plural ethos than the big guns of metropolitan cities.

The range of small papers is really wide. How many of us know that we have newspapers in 77 folk languages and dialects, in this country besides 19 principal languages. We have newspapers. not only in kashmiri, konkani,Manipuri,Nepali, Sindhi and Sanskrit but also in Arabic, Chinese, French, Italian and Burmese. Then how many of us have heared of such Indian Languages as Angika, Gangte, Kabur and Haruti. Little doubt through the small press we can discover our own country. But we tend to ignore it, and in the process we often the pulse of the people.

The government has taken a good initiative in the form of community radio. People got a platform through which they can address issues like health, education and public facilities pertaining to their area. The local people or the community can get attention of the regional the national media through this platform. The government is giving funds and grants to help the cause. It is time we should do something to sustain and nurture our diversity through the pluralism of press in the process of change.

TRADITIONAL MEDIA AND RURAL DEVELOPMENT

Traditional media means the mediums through which the cultural traits passed from generation to generation. Keeping in view their intimacy, as they are very close to the heart of the people, folk media prove to be powerful tools of communication in the rural society.

Furthermore, the traditional media is available to all and enjoyed by persons of different age groups generally at very low cost. The most important advantage of folk media is its flexibility in day to day life and it can be introduced into traditional folk from art such as wall paintings, puppet show, folk songs, dances, dramas, melas and festivals, munadi, bioscope, proverbs, riddles, story telling, tamasha, nautanki etc. Thus, the folk art forms play a prominent role in rural development.

Traditional forms of entertainment have been used to further programmes of development since 1954 after it was discovered that they could be used to present developmental message as well. The song and Drama Division was set up as an arm of the Minister of Information and Broadcasting with the responsibility of harmonizing the country's performing arts for the development of communication and it works in close cooperation with the field publicity offices which arrange various programmes.

The traditional media plays an important role in extension education. The great task of extension education is to impart a desirable controlling influence on others through the use of effective extension teaching methods. It is better to have an inventory of the locally available and familiar traditional media for possible utilization in the field of agricultural communication. A proper blending of the traditional media with the electronic media is a welcome step in this direction. besides, occasions such as agricultural fairs, farmers' days, exhibitions, fertilizer campaigns etc. can be used by the development agencies as forum do disseminate agricultural innovations through traditional media to the farming community.

The folk art forms satisfy the innate desire for self-expression and also satisfy mans need for moral instruction combined with entertainment. In contrast with the electronic media, it preserves and disseminates in a lively manner the traditions and culture of our forefathers. Indian folk forms have a generous mix of dialogue, dance, songs, clowning etc. So it is a very important method of communicating agricultural technology to the farmers and for the benefit of rural development programmes.

The important characteristics of the traditional media are (1) the accessibility of traditional media is very wide; (2) it involves more than one sense; (3) the potential for cognitive gain retention is possible but restricted; (4) potential for effective change in the attitude of the people is high; (5) there is no change in the skill of the people; (6) the maintenance of particular message is not possible; (7) interest arousal capacity is very high; (8) range of mode choice is narrow; (9) less operational abilities and skill and (10) personal cost is not relevant.

Traditional media has greatest appeal to the masses and have qualities of touching the deepest emotions of the illiterate millions. Among these puppetry is believed to be the oldest form of popular theater in India. It is important for communicating technology to the farmers in the village life, its problems and solutions. Folk theatre form like Tamasha, Nautani, Keertana or Harikatha attract the rural audiences most, so people can be educated through the mediums to bring about desirable changes in their behavior. Street play is not like theater but it attracts a large number of people. The villagers have a great fascination for their folk dances and folk songs. Melas or country fairs are synonymous with joy and gaiety and in the rural areas where life follows a hard routine, nothing is more welcome to the people than the prospect of a festival and mela. Story telling has been one of the best and most commonly used method of instruction in informal education, religious propagation, rural development etc. Riddle is also and educational device through which elders sued to communicate knowledge. Proverbs which predominate in oral civilization represent the essence of rural wisdom and knowledge. Bioscope is also a popular folk medium sued for entertainment and for propagation of information on education, agriculture etc.

The traditional medial should be an integral part of nay rural development programme, Wherever possible, it should be integrated with the modern mass media, but in all cases integration with the ongoing extension work is vital. There is big gap between the modern scientific knowledge and the knowledge possessed by the common masses. This gap is to be bridged by communicating effectively the developmental information to the rural masses. The messages communicated through the traditional media gain access to the mind through audio and video effects. The use of the eyes and the mind produces a sense of richness in meaning on the individuals. This mental reaction is both intellectual and emotional. They create a high degree of interest and make learning more permanent. The poorest man had access to his culture, expressed either in story, poem, play, song, custom, rituals or a variety of other forms of characteristics of folk culture. As these are face to face interaction between the conveyor and the recipient, there is scope for clarification of doubts and acquisition of full information. The influence on the recipient is lasting.

The employment of traditional media for effectively conveying rural developmental messages in a language and style that will be comprehended and lied by the rural people leads to spectacular results.

***

6

Agriculture and Rural Development

INTRODUCTION

Rapid urbanization and the growth of trade and industry have reduced the significance of the agriculture sector in Delhi. Agriculture and allied activities contributed about 1.71% to the Gross State Domestic Product of Delhi at current prices (1998-99). The share of agriculture and allied activities has declined sharply from 4% in 1993-94 to 1.71% in 1998-99. Only 2.61% of the total work force in Delhi was engaged in agriculture and allied activities in 1991. This is in sharp contrast to the rest of the Indian economy where more than 60% of the total work force was engaged in agriculture and related activities.

For the 70 percent of the world's poor who live in rural areas, agriculture is the main source of income and employment. But depletion and degradation of land and water pose serious challenges to producing enough food and other agricultural products to sustain livelihoods here and meet the needs of urban populations. Data presented here include measures of agricultural inputs, outputs, and productivity compiled by the UN's Food and Agriculture Organization.

Around ten to twelve thousand years ago, human began to domesticate plants and animals for food. Before this first agricultural revolution, people relied on hunting and gathering to obtain food supplies. While there are still groups of hunters and gatherers in the world, most societies have switched to agriculture. The beginnings of agriculture did not just occur in one place but appeared almost simultaneously around thc world, possibly through trail and error with different plants and animals or by long term experimentation. Between the first agricultural revolution thousands of years ago and the 17th century, agriculture remained pretty much the same.

In the seventeenth century, a second agricultural revolution took place which increased efficiency of production as well as distribution which allowed more people to move to the cities as the industrial revolution got under way. The eighteenth century's European colonies became sources of raw agricultural and mineral products for the industrializing nations.

Now, many of the countries which were once colonies of Europe, especially those in Central America, are still heavily involved in the same types of agricultural production as they were hundreds of years ago. Farming in the twentieth century has become highly technological in more developed nations with geographical technologies like GIS, GPS, and remote sensing while less developed nations continue with practices which are similar to those developed after the first agricultural revolution, thousands of years ago.

About 45% of the world's population makes their living through agriculture. The proportion of the population involved in agriculture ranges from about 2% in the United States to about 80% in some parts of Asia and Africa. There are two types of agriculture, subsistence and commercial.

There are millions of subsistence farmers in the world, those who produce only enough crops to feed their families.

Many subsistence farmers use the slash and burn or swidden agricultural method. Swidden is a technique used by about 150 to 200 million people, and is especially prevalent in Africa, Latin America, and Southeast Asia. A portion of land is cleared and burned to provide at least one and up to three years of good crops for that portion of land. Once the land can no longer be utilized, a new patch of ground is slashed and burnt for another round of crops.

Swidden is not a neat or well-organized method of agricultural production by it is effective for farmers who don't know much about irrigation, soil, and fertilization.

The second type of agriculture is commercial agriculture, where the primary purpose is to sell one's product at market. This takes place throughout the world and includes major fruit plantations in Central America as well as huge agribusiness wheat farms in the Midwestern United States.

Geographers commonly identify two major "belts" of crops in the U.S. The wheat belt is identified as crossing the Dakotas, Nebraska, Kansas, and Oklahoma. Corn, which is primarily grown to feed livestock, reaches from southern Minnesota, across Iowa, Illinois, Indiana, and Ohio.

J.H. Von Thunen developed a model in 1826 (which wasn't translated into English until 1966) for the agricultural use of land. It has been utilized by geographers since that time. His theory stated that the more perishable and heavier products would be grown closer to urban areas. By looking at the crops grown within metropolitan areas in the U.S., we can see that his theory still holds true. It is very common for perishable vegetables and fruits to be grown within metropolitan areas while less-perishable grain is predominantly produced in non-metropolitan counties.

Agriculture uses about a third of the land on the planet and occupies the lives of about two and a half billion people. It's important to understand where our food comes from.

THE IMPORTANCE OF AGRICULTURE IN DEVELOPMENT

Historical Perspectives

In this section we provide a brief historical account of some of the most significant views about the role of agriculture in development. Economist as early as the beginning of the 20th century observed that wealthier countries were characterized by a smaller portion of their output coming from agriculture and relatively less labor resources tied to the same sector. They also noted that the process of development itself was characterized by a monotonic decline in the relative importance of agriculture and the primary sector in the economy, both in terms of GDP and employment.

One of the first economists to point this out was G. B. Fisher (1939). Later, this same generalization was formalized by Kuznets (1957), showing that this secular decline of the primary sector with development can be observed both across countries and across time.

Therefore, if the process of development is characterized by a shrinking agricultural sector, should the development "recipe" then suggest policies that are biased against agriculture (in favor of other sectors of the economy) to accelerate development? Or should agricultural growth be promoted to facilitate this structural transformation? If one looks at history, and recounts the policies that developing countries implemented from the late 1950s until the 1980s (particularly in Africa and Latin America), it would seem that they followed the first strategy.

Some economists argued that agriculture plays an important role in development. One of the first arguments in favor of the role of agriculture in development was placed by Lewis (1954) who suggested that "there are large sectors of the economy where the marginal productivity of labor is negligible, zero, or even negative." Of course these labor resources are tied to the primary sector, and are a key ingredient for industrial growth, which will occur thanks to a growing labor force coming from the primary sector. Hence, the primary sector plays, although passive, an important role in development. Later, Johnston and Mellor (1961), identified some active roles that the agricultural sector performs throughout the development path: i) agriculture provides food necessary for a growing economy, as food demand, although at a decreasing rate, grows with income (Engel's Law); ii) agricultural exports generate the foreign exchange necessary to import capital goods; iii) agriculture, as the larger sector in less developed countries, is the only sector capable of generating the savings mass that the nonagricultural sector needs for capital accumulation; and iv) a growing agricultural sector creates a larger local market for the non-agricultural sector. These Johnston-Mellor linkages still remain relevant for developing economies with a large primary sector.

Johnston and Mellor were perhaps also the first to note that successful industrialization experiences are usually preceded by periods of dynamic agricultural growth. Although this does not amount to a causality link, the authors observed that countries that embark in a successful industrialization path, first experience fast agricultural expansion, fueled not by absorbing resources from the rest of the economy, but by rapid increases in productivity. The authors tell the story of Japan in the early 20th century. Many others have mentioned this feature of development; today for example, we can tell a similar story for the Asian giants of India and China, with fast industrialization preceded by fast productivity growth in the agricultural sector, i.e. the "green revolution". As a matter of fact, countries that have been able to industrialize without having first an agricultural expansion are the exception (Hong Kong, Singapore).

However this was not a widely accepted message. During the second half of the 20th century there was no widely shared optimism concerning the role of agriculture in development. Agriculture was in practice condemned mainly by two separate schools of thought. The structuralist school (particularly strong in Latin America) provided arguments against agriculture. What eventually became to be known as the Prebisch-Singer hypothesis (Prebisch (1951) and Singer (1950)), states that the commodities that developing countries (the periphery) produce and export (i.e. primary commodities) have an income elasticity of demand less than one; as opposed to the demand elasticity of the industrial goods produced by the developed countries (the center) that have income demand elasticity that is not less than unity. Therefore in the long run the price of the primary commodities exported by developing countries relative to the price of the industrial goods imported by these same countries (i.e. the barter terms of trade) is doomed to fall. Thus, specializing and exporting these primary commodities is a losing development proposition. This hypothesis was disseminated before reliable price time series were available.

Today with good statistics it is a contentious debate if this hypothesis holds or not; principally because it depends on the primary commodity and the industrial/manufactures price index used as deflator. Also, should one account for changes in quality? Recent long-term analysis (Ocampo and Parra (2003)) suggest that for most commodities the hypothesis does not hold. In the short-run most commodity prices are pro-cyclical. In the long-run some commodity real prices are non-stationary and thus move around a stable mean, while other commodities show one time falls in early 1920s and 1980s, and some other commodities show an upward trend like the relative price of meat. Therefore, the Prebisch-Singer hypothesis may apply to explain the failed development experience of any one particular country, dependent on a particular commodity, but it can not be generalized to all developing countries and to all commodities.

The second main hypothesis that condemned agriculture is related to the belief that agriculture has low potential for growth and its multiplier effect, its ability to "pull" the rest of the economy, is very low. The first part of this hypothesis can be traced back to RosensteinRodan (1943). He proposed that industrialization can be achieved by investing in several different industries separately, even if none of them alone generates enough rents to break even. This is possible because there exist economies external to the industry (i.e. increasing returns to scale) that spill-over to the rest of the economy. If there are enough industries generating this type of positive externalities, the "big push" into industrialization can be generated. Agriculture here lies in its absence, it is a sector that does not generate these economies external to the industry.

TOWARDS A NEW PERSPECTIVE OF THE ROLE OF AGRICULTURE IN DEVELOPMENT

Agriculture's Production Linkages and Development

Linkages in the agricultural sector are easy to identify: forward linkages are mainly in the agricultural and food processing industries, in the service industry with the restaurant and hotel industries and, sometimes, public schooling. Similarly the main backward linkages are with the agricultural industries that produce animal feed, with the chemical and mineral industry for purchased fertilizers, and, depending on the degree of sophistication of the agricultural sector with the financial and business services sector and the industry of machinery manufacture. In many developing countries, backward industries consist of a large number of small firms (fertilizer mixing, small scale transport, agricultural implement repair, commerce, etc.) largely labor intensive and vital for the rural economy.

Common wisdom, as reflected in Hirschman's indictment of agriculture, suggests that as the agricultural sector becomes more developed, its backward linkages increase, by requiring more financial services, machinery and other purchased inputs. Also, the forward linkages are more important in a developed economy, where there is an existing and more developed food industry, and equivalently a hotel and restaurant industry. Therefore, common wisdom suggests that in terms of linkages, agriculture would not be a good sector to promote in early stages of development, because its linkages (and thus its multiplying effect) is low. These are important questions of rural and overall development that have rarely been studied systematically.

Testing the " linkages" hypothesis The place where the linkages between sectors are recorded in an economy is the inputoutput (I-O) matrix, which shows how the total output of each sector is distributed between final consumption, from households and the government, and intermediate inputs sales; describing how each sector sells inputs to all the other sectors of the economy. In the 1950's, when there was more faith in planning, these I-O matrices where used to discover the sector with the higher backward linkages, that is the sector that could have a greater effect in "pulling" the rest of the economy. The method applied was to calculate and find the higher Leontief multiplier. When a sector expands it demands inputs from other sectors to grow; in turn, these other sectors to supply these inputs need to expand and demand more inputs from yet some other sectors. This process continues infinitely, but in each round the size of the expansion is smaller. The limit of this process of expansions generated by the growth in one sector is given by inverting a transformation of the I-O matrix, which solves for the Leontief multipliers. This method has been usually criticized because as inputs are calculated in values, and not in units, any I-O analysis assumes fixed relative prices, an unrealistic assumption when the relative size of sectors change significantly. This criticism is valid, but the I-O analysis still provides relationships that are valid at the margin.

The backward linkages, the focus of early development economists, not only are higher at earlier stages of development, but these linkages are among the highest of any sector at earlier stages of development. In the case of Bangladesh actually the highest, since Rice Milling is totally linked to agriculture, and agriculture itself is second in the ranking. Furthermore, even in middle-income countries, agriculture still has high backward linkages, as in the case of Chile, where the sector represents 4% of national GDP, but ranks 10th out of 71 sectors in importance of its backward linkages.

Understanding Structural Transformation as a General Equilibrium Process

The section above points to agriculture as a key sector to promote development as its backward linkages are highest at earlier stages of development. Therefore, a development strategy that promotes agriculture will indirectly "pull" the rest of the economy towards development. However, looking at output linkages may not be the key for the design of longterm development policies. As Timmer (2005) states, part of the controversy of the role of agriculture in development stems from the fact that structural transformation is a general equilibrium process that can not be explained by looking at agriculture alone. López et al. (2007), explain the process of reduction of the relative size of agriculture, i.e. structural change within the context of a two sector endogenous growth model. The authors show that even when labor productivity grows at the same rate in agriculture as in the industrial sector, and even if there is no declining income demand elasticity for the agricultural good, agriculture will relatively contract because one of its production assets cannot be indefinitely accumulated. This asset is of course a renewable resource, which can be understood as water, land, soil quality, etc., which is limited in its supply by nature. The income maximizing path is achieved by equating the returns to assets. The relative contraction of agriculture, and the migration of assets from agriculture into the industrial sector is a result of this equating of marginal returns to assets that can be employed in agriculture or industry, like labor.

The equation of marginal returns to assets, that at the aggregate level are always decreasing can be considered as the golden rule of development, not the promotion of the sector with high multiplier effect. While this has rather obvious policy implications and one that most economists would recommend intuitively, unfortunately if one looks at the "import substitution" era one finds the most egregious anti-agricultural bias and incredible disparities of returns to assets within the rural and urban world. For example, in China and India, the pursued industrialization policies implied a strong pro-urban bias leaving the returns to investment in the rural areas with much higher rates of return. In a very influential work, that is partially responsible for turning the anti-agricultural tide Krueger et al. (1991) (with main results summarized in Krueger et al. (1988)) show that on average (over 16 developing countries for the period 1975-1984) the exchange rate overvaluation and import tariffs on industrial goods were the equivalent to at least an 11% export tax on each country's agricultural export.

Another important lesson that can be obtained from a general equilibrium development analysis, is an interpretation of poverty traps, i.e. those countries that are unable generate enough savings to lead them into the development path. These countries are characterized by overstretching their natural resources, a manifestation of this overstretching being an excess of labor resources employed in resource using sectors. The general equilibrium view suggests two ways to exit a poverty trap: On one hand, there is implicitly low productivity in the industrial sector, and investing in non-agricultural productivity can help increase the demand for labor in the industrial sector and reduce the pressure on the resource. However, another possibility is to increase agricultural productivity to make the operation sustainable. Today most of the rural poor live in fragile tropical ecosystems, and where labor is not a major economic constraint. Most of the human and financial resources, public and private, that are devoted to agricultural R&D, are employed in research that is relevant for capital intensive and temperate climate agriculture; i.e. which are the needs of developed countries' agriculture. The corollary is that although agriculture may play a key role in development, as we have argued above, this role may not be played if the agroecological conditions places limits to the expansion of the sector.

AGRICULTURE'S PRO-POOR ROLE

Not all growth experiences are equal. There is a growing focus on the importance of a "propoor growth" defined as growth (an increase in average income / purchasing power) that is also accompanied by an improvement in the distribution of income. To help in the understanding of a formal definition of the concept, let us imagine a rural world with only 100 inhabitants. One person owns the only farm, and earns 101 monetary units, while the rest are 99 laborers that earn 1 currency unit each. If the poverty line is 2, then the headcount poverty rate in this rural economy is 99%. Let us consider 2 growth possibilities: (i) the owner earns 151.5 and the workers earn 1.5 each; and (ii) the owner earns 201 while the laborers still earn 1 each. Initially, the average income is 2, so on average nobody is poor. In both growth possibilities, income grows exactly by an striking 50%, however in both cases poverty remains at 99%. In the first case, all incomes grow exactly by 50%, so there is no change in the distribution of income, and all inequality measures commonly used (which are sensitive to the distribution and not to the level of income) remain constant. On the second case, while growth is the same, the distribution of wealth deteriorates, and income inequality measures rise.

The above example serves to highlight some important facts about growth. First, when the initial distribution is extremely unequal, growth is not enough to reduce poverty, even if there is "trickle down" of the benefits of growth, as in case (i). Given that many examples of growth that resemble cases (i) and (ii) have been documented, with limited or no reduction of poverty, is that researchers have turned their attention to "pro-poor growth". In our example assume case (iii) all incomes grow by 1. In case (iii), growth is also exactly 50%, but the distribution of income has improved, and more importantly poverty has been reduced to 0%. Case (iii) is an example of pro-poor growth.

Different formulas have been proposed to formally define pro-poor growth. Both Ravallion and Chen (2003) and Kakwani et al. (2004) propose formulas to measure "pro-poor growth" that involve scaling up or down the observed growth rate by a factor that is greater (less) than one if inequality has been reduced (increased). There is ample theoretical support and empirical evidence that suggests that agriculture is pro-poor, and that growth based on the expansion of the sector is pro-poor growth. We can identify four main channels by which agricultural growth helps poverty alleviation:

- Directly increasing the income/own consumption of small farmers: Small holders are usually not only inadequately endowed with land, but usually also lack other assets, like physical and human capital, and thus are usually poor. Expansion of the agricultural sector may benefit also the small-holder sector and pull some of them out of poverty. Small farms are, with respect to capital and land utilization, labor intensive (own family labor), and therefore are likely to benefit from technological progress that is labor intensive. When land distribution is equitable, it will be the case that expansion of agriculture will benefit the small-holder sector; when the land distribution is inequitable there could be agricultural growth fully based on large farm output expansion, in which case the small holder sector would not necessarily benefit.

- Indirectly by reducing food prices: Most measures of poverty are based, directly or indirectly in the cost of access to food. When the price of food is reduced, there is a two-way accounting improvement in the welfare of the poor. In the first place their real income increases, and more so than the wealthier, because food is the main component of their consumption basket. At the same time, the poverty line which is usually used to measure poverty is decreased, which acts to alleviate poverty. In a completely open economy without any additional transaction costs, the price of food should not be affected by agricultural growth in the same country. However, not all food is tradable, many perishable vegetables are for all purposes non-tradables. In addition, the bulk of cereal staple foods produced and consumed by the poor in rural areas are traded in local markets which, due to high transactions costs are disconnected from larger (including international) markets.
- Indirectly by increasing the income generated by the non-farm rural economy: The rural non-farm economy in most regions is either: mostly composed of goods and services that directly serve agriculture, or indirectly depend of the demand of those tied to agriculture. The more disconnected the rural economy is from urban markets, the more dependent is the rural non-farm sector on the income generated by what is usually the main engine of the rural economy: agriculture. Hence, agricultural growth can increase the demand for the goods and service of the rural non-farm sector and help pull out of poverty households tied to this sector.
- Indirectly by raising employment and wages of the unskilled: Agriculture is usually intensive in unskilled labor. Thus,

agricultural growth through an increase in unskilled labor demand will increase unskilled employment and/or the wages of the unskilled, most of which are poor. There is here a general equilibrium effect, because raising the unskilled wages in agriculture pushes upward the unskilled wages in urban areas also. This latter general equilibrium effect will be higher the more integrated rural and urban labor markets are.

Given these channels through which agriculture reduces poverty, it should come as no surprise that the overwhelming empirical evidence shows that agricultural growth is not only pro-poor, but more pro-poor than other sectors of the economy. The key in this fundamental result lies in that all four channels described benefit the poorest households of the economy more than the rest. In a seminal study Ravallion and Datt (1996) used a long time series data from India, to explain poverty with output from the different sectors of the economy. They found a large elasticity of poverty with respect to primary output (-1.2), with the sector being more effective at poverty alleviation than industry. Kakwani (1993) used the additive property of the most popular poverty indicators, the FGT (Foster, Greer, Thorbecke) class, and decomposed the effect of sectoral growth in poverty. Using the information of a 1985 Côte d'Ivoire household survey she shows that the elasticity of poverty with respect to agricultural output is much larger (-1.8) than other sectors such as services (-0.1) and industry (-0.1).

As rural economies develop, the non farm economy becomes more sophisticated and diversified, and as we show below more important within the rural economy. Therefore, this indirect effect becomes more important. On the other hand, with development, the rural non farm economy becomes more tied to the rest of the economy, diminishing this indirect effect. Thus we also expect in very poor economically isolated rural economies, the non-farm sector to depend heavily on agricultural productivity, income and demand. This is consistent with the very high demand multiplier from agriculture found by Haggblade et al. (1989) in poor Sub-Saharan rural economies; but also consistent with the high indirect poverty elasticity of agriculture found by de Janvry and Sadoulet (2002) for Latin America in general.

Finally, as agriculture makes its transformation into commercial farming, the direct effect of the sector on small and poor farmers becomes much smaller, but commercial farmers are employers of unskilled labor; and concomitantly with the commercial farm transformation and their interaction with world markets, generally rural economies become interlinked with the rest of the economy, in particular with urban markets. These are the preconditions for agriculture to have an important effect on the employment and wages of unskilled workers.

Anríquez and López (2007) show that in Chile, where the lion share of agricultural output comes from medium and large commercial farms, agriculture is still more pro-poor than other sectors of the economy, and that this poverty to agricultural growth elasticity is mostly explained (90%) by the effect of agriculture on unskilled labor markets.

LAND HOLDING PATTERN

As per Agricultural Census, 1995-96, the NCT of Delhi had a total of 34,888 agricultural holdings. This includes 19,768 individual holdings, 14,296 joint holdings and 824 institutional holdings. Out of individual and joint holdings totalling to 34,064, a majority number of 30,346 were managed by males and remaining 3,718 were managed by females. While the holdings managed by males accounted for 43,864 hectares of operated area, females managed 2,854 hectares of operated areas. Out of individual holdings totalling 19,768 a majority number of 16,774 were held by males and remaining 2,994 were held by females. While the individual holding held by males accounted for 15,991 hectares of operated area, individual female operated 1,901 hectares area.

The total area operated by all social groups in Delhi declined by 3.27% from 49,197 hectares during 1990-91 to 47,587 hectares during 1995-96. During the same period, the area operated by marginal and small holding increased by 1.20%. On the other hand, area operated by semimedium, medium and large holdings declined by 2.29%, 3.89% and 16.81% respectively. The number of holdings in all the categories has also indicated a decline. The comparative statistics on number and area of holding corroborate the trend of urbanization of Delhi and consequent conversion of agricultural land to non-agricultural uses.

LAND UTILIZATION

The total area of Delhi is 1,47,488 hectares. As per village records, the total cropped area was 60,885 hectares (41.28%) in 1999-2000.

The remainder was being used either for non-agricultural purposes or remained as forest, fallow land, uncultivable land, etc. The land-use pattern in Delhi for the period from 1995-96 to 1999-2000 (Table 10.1), reveals that the total cropped area has decreased from 62,966 hectares in 1995-96 to 48,917 hectares in 1996-97. Further total cropped area has been increased to 60,885 hectares in 1999-2000 having a growth of 24.46% as comparison to 48,917 hectares in 1996-97. In percentage terms, the total cropped area, which was 42.69% of the total available area in 1995-96, dropped to 31.17% in 1996-97 and reached 41.28% in 1999- 2000. The pace of urbanization has increased significantly in the 1990s, resulting in rapid decline in agriculture's share both in GDP and employment.

CROP INTENSITY

Crop intensity is an index of agricultural development and is directly related to irrigation facilities. The following statement shows that crop intensity in Delhi ranged from a minimum of 121% to a maximum of 147% during the period 1992-93 to 1999-2000.

Cropping Pattern

The main crops grown in Delhi are wheat, jawar, bajra and paddy. The cropping pattern in terms of production and yield per hectare for the period 1995-96 to 1999-2000 shows that both in terms of production and average yield, wheat was the predominant crop in Delhi throughout the period of five years, followed by paddy, bajra and Jawar. Wheat production constituted 87.54% of total grain production in Delhi in 1999-2000. The yield of different crops has fluctuated significantly during this five year period. Food-grain production varied from 1,10,343 tonnes in 1995-96 to 1,94,206 tonnes in 1999-2000.

The farmers of Delhi have shifted from traditional agriculture to high value commercial agriculture. The Agriculture /Horticulture Unit of Development Department is encouraging farmers to take up vegetable, floriculture production, Mushroom cultivation etc. About 4145 hectares of area has been brought under Floriculture.

Irrigated Area

The three main sources of irrigation in Delhi are canals, wells, and tubewells. The area irrigated by each source for the period from

1994-95 to 1998-99 is shown in Table 10.3. 92.55% of the net irrigated area was irrigated by wells/tubewells in 1998-99. During 1999-2000, of the total cropped area of 60,885 hectares, the net area irrigated was 39,747 hectares, area i r r igated more than once was 9321 hectares and the gross area irrigated was 49,068 hectares, which constitutes 80.59% of the total cropped area.

ANIMAL HUSBANDRY

Animal husbandry is an important component of the agriculture sector. According to the Livestock Census of 1997, the total livestock population (excluding poultry) in Delhi was 3.35 lakh. The main livestock products are milk, eggs and meat. Production estimates of these three products for the period between 1996-97 to 2000-2001 are presented in Table 10.4. Milk production has increased from 264.42 thousand tonnes in 1996-97 to 292 thousand tonnes in 2000-01 registering a growth of 10.43% during the period of five years. The total production of eggs has fallen by 38.41% from 67.41 million in 1996-97 to 41.52 million in 2000-01. Meat production has increased by 32.40% from 25383 tonnes in 1996-97 to 33609 tonnes in 1998-99 and further fallen to 33175 tonnes in 2000-2001. The number of veterinary hospitals and dispensaries in Delhi is 48 and 27 respectively. The number of animals that received treatment from the hospitals/ dispensaries increased from 3,58,000 in 1996-97 to 3,81,529 in 1998-99.

FORESTS

The area covered by forests was 10,528 hectares at the end of March 2001. This is 7.14% of Delhi's total area of 1,47,488 hectares. According to the latest report of Forest Survey of India, released in 2000, the forest cover of Delhi has increased from 26 sq.km in 1997 to 88 sq.km in 2000. To maintain the ecological balance and check envi ronmental pol lut ion, the development plans of Delhi have given high pr ior i ty to af forestat ion programmes. Plantat ion programmes have been taken up by var ious agencies on government land, pr ivate land and on land belonging to the Gram Sabha and local bodies. During 2001-2002 about 3.92 lakh trees will be planted and 1.4 lakh seedlings/ saplings distributed to various agencies and individuals. The notified forest area in Delhi is now 10528 hectares. The Forest Department of the Delhi Government has set up a Bird-cum-Wildlife Sanctuary in the Asola-Bhatti Mines area on 2,733 hectares of land.

OBJECTIVE OF THE FISHERIES

The objective of the Fisheries Unit is to regulate fishing under the Punjab Fisheries Act, 1914 and the Indian Fisheries Act, 1857, and prevent destructive practices and unscrupulous fishing. The Fisheries Unit performs the dual function of producing fish-seed as well as providing fish for consumption.

Rural Development

In 2001, the rural area in Delhi was 37.50% (556.17 sq. kms.) of the total geographical area of 1,483 sq.kms. Of the total population of 137.83 lakhs, 6.98% (9.63 lakh people) were living in the rural areas according to the 2001 Census. Over time, the extension of the urban area has converted rural villages to urbanized villages. The number of rural villages has decreased from 304 in 1951 to 165 in 2001 Census. Details of the total population of Delhi, rural population and number of villages over the period of 50 years from 1951 to 2001 are given in Table 10.7. The share of rural population declined from 18% in 1951 to 7% in 1981, but increased thereafter to 10% in 1991 and further decreased in 2001 Census to 7%. Out of the total rural population of 9,49,019 in 1991, the total workforce in rural areas was 2,72,864 (29%). There was a steep increase in the rural workforce between 1981 and 1991, from 1,28,853 in 1981 to 2,72,864 in 1991. There has been a significant shift in favour of the manufacturing Sector.

Reforms with a Human Face: Addressing the Landless Poor

Reforms in the agricultural sector are an important step toward increasing growth rates in the Indian economy and thus reducing poverty sustainably. But many households are not in a position to share in economic growth because of their low asset base (for example, poor nutrition, low education, and few physical assets).

Studies reveal that there is typically little mobility out of extreme poverty, and many households remain poor for generations. Indeed, low human capital status and an inability to build up a minimum physical asset base play a key role in the intergenerational transmission of poverty. Any credible, broad-based development strategy must therefore involve public policies aimed directly at promoting asset accumulation by chronically poor households.

In addition, the labor productivity of the poor is currently impaired by nutrition problems, including "hidden hunger" in the form of micronutrient deficiencies. Agricultural research and production programs should focus on addressing these deficiencies through supplementation, fortification of foods (including complementary foods), and attention to making low-cost foods that are rich in micronutrients.

India is home to a wide range of social safety net programs that together attempt to address the needs of poor households at various stages of the life cycle. For households with young children, the Integrated Child Development Scheme (ICDS) provides take-home food rations linked to acquiring nutrition guidance and crucial health care. To promote higher educational attainment, the Mid-Day Meals Program provides meals to children attending school. The Public Distribution System (PDS) provides subsidized food rations to poor households through a vast network of fair-price shops. A range of community public works programs (such as Jawahar Gram Samridhi Yojana or Employment Guarantee Schemes [EGSs]) provide employment to the poor during periods of economic downturn or during the slack agricultural season. The National Old-Age Pension program and the Annapurna program provide cash to destitute elderly households without alternative family support. These programs should be transformed from social assistance to social development programs that contribute directly to the creation of physical and social assets.

Although safety net programs in India vary widely and absorb substantial public funds, their combined effectiveness is questionable. Rationed food subsidies are often poorly targeted, and corruption prevents much of the food from reaching the intended beneficiaries. For example, 53 percent of India's rural poor live in three states (Bihar, Uttar Pradesh, and Madhya Pradesh), but their dependence on subsidized food through the PDS off-take is only between 5 and 10 percent of their total cereal consumption—too little to make much difference in their food security. The costs associated with public distribution of food are also often unnecessarily high.

There is a need to rationalize wages in public works programs, walking the line between too-high wages, which will result in leakage of transfers to nonpoor households, and too-low wages, which will undermine the very objective of the programs— that is, poverty alleviation. These two considerations need to be balanced, in line with minimum wage regulation. In addition, high costs associated with managing the creation of assets through public works programs absorb scarce resources, and the resulting projects are often of low quality or never benefit the poor. The economic inefficiencies associated with financing these safety net and public works programs can also be substantial, as is the case with foodgrain support prices that distort production incentives.

These different safety net programs are often poorly integrated, with some households receiving benefits from a number of sources and other poor households being completely excluded. As a first step, existing social safety net programs in India need to be revisited to assess their targeting mechanisms, coverage, cost-effectiveness, and overall impact on poverty alleviation. Research at IFPRI, along with several studies in India, shows that programs like the EGS of Maharashtra to build rural infrastructure are more cost-effective in reaching the poor than is the untargeted PDS. These public works schemes need to be scaled up to build rural infrastructure, develop and preserve watersheds, undertake forestation, desilt canals, and so forth. Bangladesh's Food for Education (FFE) scheme and India's own ICDS show that targeted programs have been highly successful and are worth investigating. Under the FFE scheme, the poor family of the school-aged child gets a quantity of subsidized food as long as the child attends school. This program ensures higher attendance in village schools, especially of girls, and provides food security to the poor. Such a program may be worth implementing in India on a pilot basis. Achieving reforms for the landless poor requires developing and applying credible evaluation techniques that can then inform the design and implementation of programs. Given budget constraints and the extent of poverty in India, the country cannot afford to tackle the problem of assisting the landless poor without substantial improvements in the cost-effectiveness of the overall social safety net system. Fortunately, we have learned much from diverse experiences in several developing countries, and the wide variation in program performance across Indian states may also be a valuable source of lessons for future policy reforms.

Toward High-Value Agriculture

Given sustained increases in per capita incomes of about 4 percent per year during the past two decades, consumption patterns in India are changing away from cereals to high-value agricultural products. How fast has the consumption basket of an average Indian changed? Data from the National Sample Survey Organisation (NSSO) show that per capita consumption of cereals from 1977 to 1999, for example, declined from 192 to 152 kilograms per year in rural areas and from 147 to 125 kilograms in urban areas. The consumption of fruits, on the other hand, increased by 553 percent, of vegetables by 167 percent, of milk and milk products by 105 percent, and of meat, eggs, and fish by 85 percent in rural areas over the same period. Similar changes occurred in urban diets. These dramatic changes indicate a structural shift in Indian diets. Add to this the new export market opportunities for many of the same products, owing to trade liberalization, and there is a happy match between the demands of the market and the need for farmers to diversify into higher-value activities. Further, high-value agricultural products have higher employment elasticity and can be suitable for smallholders, if they can participate.

In this new situation, more of the energies and resources of the agricultural sector can be unleashed to produce the kinds of high-value foods and products that are now in high demand by India's growing middle classes and urban dwellers and that have new export market opportunities. A reinvigorated agricultural and agribusiness sector could thus continue to be a major engine of income and employment growth for the country. Despite the tremendous opportunities ahead, success is not yet assured.

Important Challenges will Need to be Overcome

The first challenge is to further shift the government's priorities from heavy support and protection of food staples to promotion of agricultural diversification, processing, and commercialization. Simply put, most farmers are not going to get rich by growing cereals when there are already national surpluses, demand growth is slow, and world markets are glutted with the subsidized production of rich-country farmers. Farmers must shift into higher-value products to increase their incomes. A set of public policies and investments is required to fully unleash this new potential. This set must include additional public investment in the kinds of rural infrastructure and technologies needed for these new high-value activities, improvements in marketing and distribution systems for higher-value and more perishable foods, and further liberalization of the agroindustrial sector. The private business sector can and should play a dominant role in these higher-value market chains, and public policy must strengthen the enabling environment. This change will require a fundamental shift in thinking in many public agencies that are still geared toward the dominant role that the state played in the market chains for food staples during the Green Revolution era.

Although some of the funding for these new investments will come from the private sector, new public investments are also needed. The needed funds might be obtained by reducing some of the huge subsidies that are still maintained on fertilizers, credit, and water for the food staples sector and that no longer serve a useful purpose. This could be a win-win strategy for farmers and the government and at the same time could contribute to national economic growth.

The second challenge for the "new" high-value agriculture is to make it pro-poor. Left to market forces alone, the major beneficiaries of the new high-value agriculture will be mostly the larger and commercially oriented farms, as well as farms that are well connected to roads and markets. The majority of the 300 million or so poor people in India are rural people who depend on agriculture for their living, and many live in the less-favored regions. These people must not get left further behind during the next phase of India's agricultural development.

Fortunately, there is great opportunity to guide the new highvalue agriculture so that small farms and even many less-favored regions can be major participants. Achieving broad participation will require improving infrastructure and education in many lessfavored regions and communities, ensuring that small farms get the technologies and key inputs they need, and promoting producer marketing organizations that can link small farmers to the new market chains (supermarkets, contractors, processors, exporters, and the like). Small farmers cannot do all of these things on their own, and the public sector, private sector, and NGOs all have important roles to play. Because high-value agriculture demands more working capital, which small farmers often lack, a major effort must be made to reform the rural credit delivery system to reach smallholders. Innovative institutions promoting vertical coordination between farms, firms, and forks (supermarkets) would reduce transaction costs and market risks and would also act as a conduit to funnel more credit into this venture, especially for smallholders. This system would help lay a foundation for globally competitive agriculture in which smallholders can also participate and prosper. Public policy can make a major contribution by facilitating farmer organizations, standardization, transparent food safety policies, and contract security between farmers and the processing and retail industry.

A third challenge will be overcoming many of the environmental problems that now plague agriculture. Water scarcities will continue to grow, and farmers must learn to use less water and to be less polluting. Land degradation and deforestation must also be contained. A shift toward more diversified and higher-value farming systems will help, both because many of the new crops need less water and because, by increasing returns to land, small farmers will have less need to overexploit poor lands and soils.

Although agriculture can make a significant contribution to growth, employment creation, and poverty reduction, on its own it will not drive the full economic transformation that is now possible for India. A fourth challenge, therefore, is for policymakers to find ways of accelerating growth in the service and manufacturing sectors, which will require continued economic liberalization and privatization.

CONCLUSION

We suggest five areas for action to put rural India on a higher growth trajectory that would cut hunger, malnutrition, and unemployment at a much faster pace than has been the case so far. The five areas for action are interlinked and would best work if pursued in conjunction. We emphasize investments with a human face that include and reach out to the rural poor and a reorientation of subsidies toward such investments.

1. India should increase investments in rural infrastructure (including transport and information technology that connects villages) and agricultural R&D (leading to

improved technologies for farmers). This is our most important suggestion. To ensure high returns on these investments, India will have to invest in institutions that make implementing agencies transparent and accountable to user groups. Part of this expansion of pro-poor investments in rural India should be financed by reducing food and input subsidies, making them available only to vulnerable groups.

2. India should reorient its social safety nets to create more employment in rural areas; help strengthen the human resource base through education, nutrition, and empowerment of women; and build physical infrastructure. In this context, schemes like the EGS of Maharashtra to build rural infrastructure and FFE, well tested in Bangladesh, are much more promising than the untargeted PDS. These social investments must also address the high prevalence of micronutrient deficiencies (especially of iron, vitamin A, and zinc) among the poor.
3. Water is going to be increasingly scarce. Investing large sums in new mega-irrigation schemes may not be the best course of action, but it is important to complete those in which a lot of money has already been invested. Overall, however, managing water use through institutional changes, such as water rights that are based on farmer groups and water-harvesting schemes in dry areas with local participation, are likely to be more rewarding. Price reforms in irrigation, and even power supplies for agriculture, can succeed only if accompanied by suitable institutional reforms.
4. Indian agriculture faces promising opportunities in the production and marketing of high-value livestock products, fruits and vegetables, and fishery. To exploit these opportunities, India must liberalize its marketing and trade policies to encourage vertical coordination between farms, firms, and forks (supermarkets); facilitate increased flow of rural credit, especially to smallholders, through, say, nonbanking financial intermediaries; and withdraw any special concessions in support of foodgrain policies.

5. Trade liberalization in agriculture has the potential to bring rich dividends to developing countries, including India. To realize this potential, India must work toward establishing and strengthening a rules-based multilateral trading system through WTO negotiations. In the event of major hurdles in WTO negotiations and a delay in reaching any substantive agreement, India should explore its second-best options of reaching bilateral or regional free trade agreements with major developing countries in the region and beyond. Furthermore, to exploit the full potential of trade liberalization, India should carry out "behind the border" reforms by streamlining its own domestic markets, institutions, and infrastructure.

***

7

Rural Society

INTRODUCTION

Rural society is a society in which there is a low ratio of inhabitants to open land and in which the most important economic activities are the production of foodstuffs, fibres, and raw materials. Such areas are difficult to define with greater precision, for, although in nonindustrialized nations the transition from city to countryside is usually abrupt, it is gradual in industrialized societies, making it difficult to pinpoint the boundaries of rural places. A second, related problem is that governments do not use the same statistical criteria for rural and urban populations; in Japan, for instance, any cluster of fewer than 30,000 people is considered rural, whereas in Albania a group of more than 400 inhabitants is regarded as an urban population.

In the past, rural societies were typified by their adherence to farming as a way of life. Such cultures were not goal- or achievement-oriented; their members sought subsistence, not surplus. Marked by a high regard for intimacy and traditional values, farming communities were often regulated by kinship customs and ritual, and, in particular, the ownership and care of productive land was strictly guarded by tradition. Collectively, these characteristics are often designated by the term gemeinschaft, an expression introduced by the German sociologist Ferdinand **Tönnies. Tönnies** described the contrasting nature of urban life with the term gesellschaft, a state characterized by impersonal bureaucracy, rationalized specialization, and mechanization. **Gesellschaft** is typically associated with modern industry, where people are employees who perform specific, goal-oriented functions in a rational and efficient, as opposed to a traditional and organic, manner. The two terms are sometimes translated as "community" and "society." Rural inhabitants work with people they know well and are accustomed to relationships of great intimacy and small scale, whereas urban dwellers know each other in narrow, segmented ways that have little to do with family or friendship. According to Tönnies and subsequent sociologists, all societies are characterized by mixtures of gemeinschaft and gesellschaft; the United States, where even agriculture is almost completely mechanized, is closer to the gesellschaft end of the spectrum, whereas rural India, which is still heavily guided by tradition, is an example of gemeinschaft.

Historically, farming societies have had higher birthrates than urban societies; their populations have also tended to be younger, to live in larger families, and to include slightly greater percentages of males. These phenomena were related: it was to a farmer's advantage to have many offspring, especially males, who could work in the fields as children and then would support their parents as they grew older. Generally, however, as the children became older, there was not enough productive land for all of them to support their own families, and some would migrate to the cities. In this way, cities have historically absorbed the excess population of the countryside, thus tending to become filled with comparatively older people living in smaller families. With the advent of improved health care in this century, infant mortality rates fell, and the increased number of surviving offspring has swelled the number of migrants to the cities.

In the industrialized nations the countryside has sometimes been virtually depopulated, to the point that, for example, in 1970 only 6.7 percent of the employed persons in the United States were in the fields of agriculture, fisheries, and forestry. The result has been a global acceleration of the process ofurbanization, which has in turn created vast slums in many urban centres. To halt or reverse this process, agricultural-development specialists have suggested methods of increasing productivity without moving large numbers of farmworkers off the land. Among their recommendations are improvements in soil technology and changes in irrigation, seed stocks, and drainage; they counsel against further large-scale mechanization. The habit of the developed nations to apply their own practices of agriculture to situations where they might not be ultimately beneficial has been yielding to the belief that appropriate technologies must be developed for each area.

EFFECT OF MOTOR VEHICLES IN AUTOMOTIVE INDUSTRY: SOCIAL EFFECTS

Economic and Social Significance

The automotive industry has become a vital element in the economy of the industrialized countries—motor vehicle production and sales are one of the major indexes of the state of the economy in those countries. For such countries as the United Kingdom, Japan, France, Italy, Sweden, Germany, and South Korea, motor vehicle exports are essential to the maintenance of healthy international trade balances.

The effect of motor vehicle manufacturing on other industries is very great. Almost one-fifth of American steel production and nearly three-fifths of its rubberoutput go to the automotive industry, which is also the largest single consumer of machine tools. Moreover, the special requirements of automotive mass production have had a profound influence on the design and development of highly specialized machine tools and have stimulated technological advances in petroleum refining, steelmaking, paint and plate-glass manufacturing, and other industrial processes.

The indirect effects are also considerable through the many auto-related businesses, such as motor freight operators and highway construction firms. In addition, truck transportation has grown steadily throughout the world.

Highway Development

Before the advent of the motor vehicle, roads in most parts of the world were generally poor. The available methods of road transport were so costly and inefficient that, unless there were special considerations such as military movements, it was not worthwhile to maintain roads for other than local traffic. The general use of automobiles created a strong demand for better highways. The first response was to provide for the improvement of existing road networks. Experience subsequently demonstrated that roads for automobile traffic needed to be differentiated functionally, depending on whether they were intended for through traffic or local traffic. Main arteries are best designed as freeways (motorways, *autostrade*, or *Autobahnen*)—i.e., divided highways with complete control of access and no intersections at grade.

Social Effects

A historian has said that Henry Ford freed common people from the limitations of their geography. The statement cogently summarizes the social transformations still proceeding throughout the world as a result of the motor vehicle. It has created mobility on a scale never known before, and the total effect on living habits and social customs is still incalculable.

The automobile has radically changed urban life by accelerating the outward expansion of population into the suburbs and beyond. As with other automobile-related phenomena, the trend is most conspicuous in the United States but is rapidly appearing elsewhere. The decentralizing trend is accentuated by the fact that highway transportation encourages business and industry to move outward to sites where land is cheaper, where access by car and truck is easier than in crowded cities, and where space is available for the one-story structures that permit optimum use of modern materials-handling techniques. Yet the effect on rural life has been, if anything, more pronounced than the effect on cities. In the days of horse-drawn transport, the economical limit of wagon transportation was 15–25 km (about 10–15 miles); any community or individual farm more than 25 km from a railroad or navigable waterway was isolated from the mainstream of economic and social life. Motor vehicles and paved roads have narrowed much of the gap between rural and urban life. Farmers can ship easily and economically by truck and can drive to town when convenient. In addition, such institutions as regional schools and hospitals are now accessible by bus and car.

It would be impossible to list all of the specific effects of motor vehicle production, but two are especially illustrative. First, the marketing of automobiles has stimulated a great expansion in the use of credit. Installment buying existed before the automobile but in a limited scope. The technique was introduced into the American automobile industry in 1916 by manufacturers of medium-priced cars to help meet the competition of the low-priced Model T. It became a universal practice in nearly all countries in the purchase of motor vehicles, and it accustomed people to buying other durable consumer goods in the same way. Second, there has been a striking development of businesses such as drive-in and drive-through eating establishments and of commercial developments, such as shopping malls, that are designed to be accessed primarily by car.

In both urban and rural areas after World War II, the automobile is credited with having caused drastic changes in the sexual values of young people, who found in it a privacy not formerly attainable.

Recreational Travel

One of the conspicuous effects of the automobile has been to permit nearly everyone in the automotive countries to travel for recreation. The motor vehicle allows for such auxiliary devices as trailers (called caravans in Europe), campers, trailers for boats and off-road vehicles, and bicycle and ski racks, which broaden the scope of recreational opportunities.

Adverse Effects

The mass use of motor vehicles was bound to have some unforeseen and undesirable consequences, of which three can be singled out: traffic congestion,air pollution, and highway accidents. The approach to each of these problems illustrates a common propensity to blame the technology, rather than the way in which the technology has been used.

City streets were congested long before the automobile existed, but the problem has been compounded enormously by the masses of motor vehicles that enter or leave cities at peak traffic hours. The constantly growing number of automobiles throughout the world adds to the difficulty of finding remedies for congestion. The heart of the problem is that few city street systems have been designed for automobile traffic. Reliable estimates are that some two-thirds of the vehicles in central business districts are only passing through and should have been routed on circumferential highways. Remedying this situation is difficult and expensive. It calls for modern highways to provide both ready access into downtown areas and ways to avoid them. Programs for this purpose encounter vigorous opposition, frequently justified, on the ground that building freeways in cities disrupts neighbourhoods and destroys scenic or historic areas.

The widespread use of automobiles for business travel has also led in many cities to a decline in public transit systems, and the need to develop and use mass transit has been much discussed. Given the trend toward dispersal of people and businesses in urban areas, it seems doubtful that mass transit will appreciably diminish motor vehicle traffic. Still, in most cities, bus systems can provide the needed capacity for public transportation and are the most economical way of doing so.

Atmospheric pollution antedates the automobile, but the concentration of many thousands of motor vehicles in large cities has given the problem a new dimension. Automobile exhaust commonly contributes half the atmospheric pollutants in large cities and even more in cities where atmospheric and topographic conditions are peculiarly conducive to smog formation. In the 1960s federal and state legislation in the United States required the installation of controls on motor vehicles to restrict the emission of pollutants. At the end of the 20th century, many scientists believed that emissions from motor vehicles, industrial processes, and power plants were leading to a buildup of carbon dioxide in the atmosphere, thus trapping additional heat and raising Earth's temperature with potentially disastrous long-term results. Although not universally accepted, the theory of led governments in many major automotive countries to enact legislation requiring a significant increase in motor vehicle fuel economy, thereby reducing the output of carbon dioxide. Many automobile manufacturers also have undertaken development of alternative, less-polluting power sources, such as fuel cells that convert hydrogen (derived from gasoline, natural gas, methanol, or other sources) and oxygen into electricity to power an electric motor, to enhance their competitive positions even in countries without strong requirements that they do so.

Highway accidents create a distressing toll of fatalities and injuries wherever there is widespread use of automobiles. Each year there are hundreds of thousands of motor vehicle fatalities worldwide and more than 40,000 in the United States alone. The social and economic cost of such accidents is incalculable. Efforts to improve highway safety have been successful in most countries, but a reduction in the ratio of fatalities and injuries per distance traveled is often offset by increases in numbers of accidents because of the ever-growing use of motor vehicles.

Safety features such as seat belts and air bags that inflate on impact have become standard features in cars and passenger trucks since the 1960s. Today many vehicles are equipped with multiple air bags to protect occupants in side-impact and rollover accidents as well as frontal crashes.

The desire to reduce fatalities and to conserve fuel has led policy makers to focus on speed limits. Most countries of the world have set speed limits ranging from about 65 km (40 miles) per hour in some island nations to 120–130 km (75–80 miles) per hour in many European countries. In some parts of the world, such as areas of Germany, India, and the Philippines, speed limits traditionally are not prescribed. In 2000, British policy makers and government researchers debated a controversial proposal to equip new cars with a speed-control device that would use global positioning satellites to track a vehicle's location and, in conjunction with an onboard digital road map, cut off the car's fuel supply if local speed limits were exceeded.

PAVEMENTS IN ROADS AND HIGHWAYS (TRANSPORTATION)

In a typical rural pavement (as shown in the figure), the top layer of the pavement is the wearing course. Made of compacted stone, asphalt, or concrete, the wearing course directly supports the vehicle, provides a surface of sufficient smoothness and traction, and protects the base course and natural formation from excessive amounts of water. The base course provides the required supplement to the strength, stiffness, and durability of the natural formation. Its thickness ranges from 4 inches (10 centimetres) for very light traffic and a good natural formation to more than 40 inches (100 centimetres) for heavy traffic and a poor natural formation. The subbase is a protective layer and temporary working platform sometimes placed between the base course and the natural formation.

Pavements are called either flexible or rigid, according to their relative flexural stiffness. Flexible pavements have base courses of broken stone pieces either compacted into place in the style of McAdam or glued together with bitumen to form asphalt. In order to maintain workability, the stones are usually less than 1.5 inches in size and often less than 1 inch. Initially the bitumen must be heated to temperatures of 300°–400° F (150°–200° C) in order to make it fluid enough to mix with the stone. At the road site a paving machine places the hot mix in layers about twice the thickness of the stone size. The layers are then thoroughly rolled before the mix cools and solidifies. In order to avoid the expense of heating, increasing use has been made of bitumen emulsions or cutbacks, in which the bitumen binder is either treated with an emulsifier or thinned with a lighter petroleum fraction that evaporates after rolling. These treatments allow asphalts to be mixed and placed at ambient temperatures.

The surface course of a flexible pavement protects the underlying base course from traffic and water while also providing adequate tire friction, generating minimal noise in urban areas, and giving suitable light reflectance for night-time driving. Such surfaces are provided either by a bituminous film coated with stone (called a spray-and-chip seal) or by a thin asphalt layer. The spray-and-chip seal is used over McAdam-style base courses for light to moderate traffic volumes or to rehabilitate existing asphalt surfaces. It is relatively cheap, effective, and impermeable and lasts about 10 years. Its main disadvantage is its high noise generation. Maintenance usually involves further spray coating with a surface dressing of bitumen. Asphalt surfacing is used with higher traffic volumes or in urban areas. Surfacing asphalt commonly contains smaller and more wear-resistant stones than the base course and employs relatively more bitumen. It is better able to resist horizontal forces and produces less noise than a spray-and-chip seal.

Rigid pavements are made of portland cement concrete. The concrete slab ranges in thickness from 6 to 14 inches. It is laid by a paving machine, often on a supporting layer that prevents the pressure caused by traffic from pumping water and natural formation material to the surface through joints and cracks. Concrete shrinks as it hardens, and this shrinkage is resisted by friction from the underlying layer, causing cracks to appear in the concrete. Cracking is usually controlled by adding steel reinforcement in order to enhance the tensile strength of the pavement and ensure that any cracking is fine and uniformly distributed. Transverse joints are sometimes also used for this purpose. Longitudinal joints are used at the edge of the construction run when the whole carriageway cannot be cast in one pass of the paving machine.

In places where the local natural material is substandard for use as a base course, it can be "stabilized" with relatively small quantities of lime, portland cement, pozzolana, or bitumen. The strength and stiffness of the mix are increased by the surface reactivity of the additive, which also reduces the material's permeability and hence its susceptibility to water. Special machines distribute the stabilizer into the upper 8 to 20 inches of soil.

In deciding whether to use a flexible, rigid, or stabilized pavement, engineers take into account lifetime cost, riding characteristics, traffic disruptions due to maintenance, ease and cost of repair, and the effect of climatic conditions. Often there is little to choose between rigid and flexible pavements.

The properties of the base course material are usually determined by laboratory tests, although field tests are sometimes conducted to check that the construction process has achieved the designer's intent. Designers typically consider the possibility of structural failure resulting from a single overload and also from damage accumulating under the passage of many routine loads. Both of these types of failure are almost entirely caused by trucks.

GEOGRAPHICAL DISTRIBUTION AND URBANIZATION

It goes without saying that populations are scattered across space. The typical measure of population in relation to land area, that of population density, is often a meaningless one, since different areas vary considerably in their value for agricultural or other human purposes. Moreover, a high population density in an agrarian society, dependent upon agriculture for its sustenance, is likely to be a severer constraint upon human welfare than would the same density in a highly industrialized society, in which the bulk of national product is not of agricultural origin.

Also of significance in terms of geographical distribution is the division between rural and urban areas. For many decades there has been a nearly universal flow of populations from rural into urban areas. While definitions of urban areas differ from country to country and region to region, the most highly urbanized societies in the world are those of western and northern Europe, Australia, New Zealand, temperate South America, and North America; in all of these the fraction of the population living in urban areas exceeds 75 percent, and it has reached 85 percent in West Germany. An intermediate stage of urbanization exists in the countries making up much of tropical Latin America, where 50 to 65 percent of the population lives in cities. Finally, in many of the developing countries of Asia and Africa the urbanization process has only recently begun, and it is not uncommon to find less than one-third of the population living in urban areas.

The rapidity of urbanization in some countries is quite astonishing. The population of Mexico City in 1960 was around 5,000,000; it was estimated to be about 17,000,000 in 1985 and was projected to reach 26,000,000 to 31,000,000 by 2000. A rule of thumb for much of the developing world is that the rate of growth of urban areas is twice that of the population as a whole. Thus in a population growing 3 percent annually (doubling in about 23.1 years), it is likely that the urban growth rate is at least 6 percent annually (doubling in about 11.6 years).

RURAL SETTLEMENT IN ASIA

Population densities have everywhere increased, and the modernization of agriculture, increased mineral exploitation, and industrialization have brought cultural change. Some of the small ethnic groups have been dying out, but larger groups often have accepted change and have increased in numbers. In South and East Asia, growing lowland populations have been pressing hard on the available land as population densities exceed 2,000 persons per square mile (750 per square km).. In Indonesia, government programs have encouraged farmers to relocate from Java, one of the most densely populated places on Earth, to more thinly populated Indonesian islands, where ethnic Javanese have sometimes come into conflict with indigenous peoples.

Similarly, in Central Asia, both Chinese and Russian settlement programs have moved peoples from heavily populated regions into frontier zones in order to develop both agricultural and industrial resources. In southern Siberia the Soviet settlement program spread a thick wedge of European Russians and assorted ethnic minorities eastward to the Pacific Ocean and northward along every river valley to the Arctic Ocean. As a result, many of the Paleo-Siberian ethnic groups have been submerged and absorbed. Old trading posts, oasis towns, and the few old cities of southern Siberia and the Central Asian republics have been developed into modern industrial centres; these have been linked to modern transport systems by which raw materials and manufactured products flow to the European regions. Most new cities have been populated largely by European Russians, with Asian peoples remaining chiefly in the rural areas. The modernization of Southwest Asia—through the renaissance of Turkey and the impact of petroleum exploitation on the Arabian Peninsula, Iraq, and Iran—has altered many of the old patterns of ethnic groupings in those areas. A further alteration of the historic pattern came in 1948 with the creation of the State of Israel, to which large numbers of Jews from North Africa, the Middle East, Europe, and North America have migrated.

RURAL SOCIETY IN FRANCE

Rural life changed more gradually. The expanding markets favoured well-endowed or efficient lords or peasants who could produce a surplus of goods for sale. Such conditions were less common in the south than in the north, although they could be found in most wine-producing areas. But, while rising prices benefited producers, they contributed to certain difficulties in the countryside. Fixed revenues in coin proved an unsatisfactory alternative to payments in kind, which landlords specified when new land was put under cultivation. Moreover, needs and tastes became more expensive and tended to exceed aristocratic resources; lavish generosity continued to be an admired and practiced virtue, and costly Crusades—occasionally lapsing into speculative adventures—regularly attracted noblemen after the end of the 11th century. Larger lordships began to employ salaried estate managers, while in the south the division of landed fortunes among numerous heirs resulted in a multiplied and impoverished petty nobility. Many rural landlords fell into debt in the 13th century. And, as wealth and nobility became less correlated, some nobles, especially those who were financially hard-pressed, sought to close ranks against the intrusion of new men or creditors. They insisted on noble birth as a condition for knighthood, reserving the designation of "squire" (or donzel, in the south) for those of noble birth awaiting or postponing the expensive dubbing (adoubement). At the upper extreme, a noble elite, the barons, achieved recognition in administration and law.

Peasant societies also became stratified. Men unable to set aside a surplus against times of famine and those who had to borrow or rent their tools or teams found it difficult to avoid dependence on other men. In some areas serfdom was renewed, or confirmed, as jurists interpreted the more stringent types of peasant obligation in the light of the revived Roman law of slavery. But here again economic and legal status did not necessarily coincide. Rich peasants who employed other men to drive their teams could be found in any village; such people as the mayor, the lord's provost, and the peasant creditor established themselves as a rural elite, whose resources insured them against calamity and opened up diverse opportunities in prospering regional economies. Where enfranchisement occurred, the lord usually received a good payment; even when servility persisted, there was a tendency to commute the arbitrary tallage into fixed common sums. New villages continued to be established, especially in the south, where many previously existent communities of peasants also received charters of elementary liberties in the 13th and early 14th centuries.

These conditions notwithstanding, the manor, or seigneurie, resisted fragmentation. The favourable market for grain and the psychological attachment of lords to their fathers' possessions preserved demesne land (for use by leasehold, not freehold, tenants) as the chief source of seigneurial income through the 13th century. The lords also continued to require the services of labourers, although the shortfall increased between work owed and work needed. Accordingly, lords resorted to paid seasonal labour, so that the margin between profit and loss became a more critical calculation than in the past. A new alternative was to lease the demesne to paid managers or sharecroppers, but this practice spread more slowly in France than in neighbouring countries. Whether lords had demesnes and servile tenants or not, the association between landlordship and power remained close. Tenancies or properties smaller than the grand old residences known as manses appeared everywhere but especially in the north, where horsepower and three-field crop rotations were making possible more productive agriculture. The burgeoning viticultures of Burgundy and Gascony proved incompatible with traditional demesne lordship and encouraged sharecropping and peasant initiative. Innovation was less common in the uplands of the centre and south, where the manse tended to retain its identity and fiscal utility.

RURAL SETTLEMENT IN GERMANY

The most striking feature of the rural settlement pattern in western Germany is probably the concentration of farmyards into extremely large villages, known as Haufendörfer. These villages are surrounded by unenclosed fields divided into often hundreds of striplike units. The Haufendorf is particularly characteristic of Hessen and southwestern Germany, areas that have a tradition of partible inheritance. During periods of population pressure, land holdings—as well as farmhouses and farmyards—were repeatedly divided on inheritance, becoming progressively smaller and more fragmented. As a result, villages became increasingly huddled and chaotic. In areas with a tradition of undivided inheritance (e.g., Bavaria and Lower Saxony), the holdings—the individual field parcels and the farmhouses—remained larger.

During the period of high medieval prosperity in the 11th to 13th centuries, population pressure brought about the advance of peasant settlements into the forests, where isolated farms and hamlets were the usual settlement form, and the colonization of the predominantly Slavic lands beyond the line of the Elbe and Saale rivers. Because this was an organized colonization under the control of the lords and their agents, the haphazard structure of the western Haufendorf could be streamlined into a few well-planned forms. Thus farmhouses in the eastern regions were customarily arranged along either a single village street (Strassendorf) or an elongated green, on which stood the church (Angerdorf); long unfenced strips of land were allotted at right angles to the road or green.

The evolution of rural settlement has not been uniform. Instead, there have been phases of advance and retreat. In particular, the decline of medieval prosperity, accompanied by the Black Death that decimated much of Europe's population in the 14th century, led to a stage of retreat, in which many hundreds of villages—the so-called "lost villages"—were abandoned in western Germany. In eastern Germany lords often appropriated deserted farms and added them to their own land, thus initiating that characteristic feature of the area beyond the Elbe, the large Junker estate farm (Gut).

The organization of agriculture and settlement in eastern and western Germany diverged considerably after World War II. In western Germany federal and state governments provided large subsidies to improve the existing structure. Land consolidation created larger holdings, and in some places farmers were moved to new farmsteads dispersed outside the villages. An increased average size of holding was associated with a massive movement of people out of agriculture. But instead of leaving for the cities, as occurred in the 19th century, people mostly remained in their existing homes and commuted to work. Part-time "Sunday" farming remained, but on a reduced scale. Land was actually left uncultivated by the new urban workers (social fallow). Some of this land was cultivated by the few remaining full-time farmers, but marginal areas were abandoned, being either afforested or reverting to rough grass and scrub.

In East Germany the Junker estates were confiscated and either divided among peasants or turned into state farms. This development was only the first stage in a process of collectivization; from 1958 to 1960, private holdings were regrouped under heavy political pressure into vast "cooperative" farms. New buildings marked the introduction of mechanized cultivation or large-scale animal husbandry, and multistory apartments and community centres reflected a politically inspired attempt to create a new concept of rural life. After unification rural settlement patterns and agriculture were once again transformed in eastern Germany, with a decline of about three-fourths in agricultural employment. Few private farms were reestablished, however, and very large areas were fallowed.

Rural Settlement in Southeast Asia

Southeast Asia is predominantly rural: three-fourths of the people live in nonurban areas. Moreover, population is heavily clustered in fertile river valleys and especially in delta areas, such as those of the Mekong and Irrawaddy rivers. Historical, cultural, and environmental influences also have affected the settlement patterns. Java and other core areas such as the Bangkok (Thailand), Hanoi, and Manilametropolitan areas contain high population densities. While the rate of urbanization in Southeast Asia is relatively low compared with those of other developing regions, it is increasing rapidly. Singapore is unique in that it is essentially totally urban. In addition, the Philippines has a much higher than average level of urbanization, in part because of its Spanish and American colonial history. The largest cities—Jakarta (Indonesia), Bangkok, and Manila—are among the world's most populous. The growth of cities of all sizes is being fueled primarily by natural increase, but rural-urban migration also is a significant contributor. Rural dwellers continue to be attracted by the promise of employment and other opportunities, but for many migrants the informal (undocumented) economic sector in these large cities is the only hope for some form of employment.

Settlement patterns in rural areas tend to be associated with agricultural practices. Shifting cultivation is still common in some parts of the region (notably the remote interior areas of Myanmar, Vietnam, and the island of Borneo), although the amount of land so utilized is gradually shrinking. The village is the unit of settlement and often functions collectively, and typically it is moved from time to time. By contrast, wet-rice cultivation, the dominant form of agriculture in Southeast Asia, is sedentary and results in relatively large rural agglomerations with well-developed village life and customs. Dry and upland farming often produces scattered homesteads.

Population resettlement to provide agricultural employment and access to land is important in some Southeast Asian countries, notably Indonesia, Malaysia, and Vietnam. By far the largest program has been conducted in Indonesia, where more than four million people have been voluntarily resettled from Java and Bali to the less populated islands. Despite considerable success, the program has been plagued by such problems as improper site selection, environmental deterioration, migrant adjustment, land conflicts, and inadequate financing. A program in Malaysia also has been quite successful, in part because it has set much smaller resettlement targets and has been better funded. Vietnamese development policy also has utilized the resettlement of people in an effort to revitalize areas outside the major population centres.

RURAL SETTLEMENT IN US

Patterns of rural settlement indicate much about the history, economy, society, and minds of those who created them as well as about the land itself. The essential design of rural activity in the United States bears a strong family resemblance to that of other neo-European lands, such as Canada, Australia, New Zealand, South Africa, Argentina, or tsarist Siberia—places that have undergone rapid occupation and exploitation by immigrants intent upon short-term development and enrichment. In all such areas, under novel social and political conditions and with a relative abundance of territory and physical resources, ideas and institutions derived from a relatively stable medieval or early modern Europe have undergone major transformation. Further, these are nonpeasant countrysides, alike in having failed to achieve the intimate symbiosis of people and habitat, the humanized rural landscapes characteristic of many relatively dense, stable, earthbound communities in parts of Asia, Africa, Europe, and Latin America.

Early models of land allocation

From the beginning the prevalent official policy of the British (except between 1763 and 1776) and then of the U.S. government was to promote agricultural and other settlement—to push the frontier westward as fast as physical and economic conditions permitted. The British crown's grants of large, often vaguely specified tracts to individual proprietors or companies enabled the grantees to draw settlers by the sale or lease of land at attractive prices or even by outright gift.

Of the numerous attempts at group colonization, the most notable effort was the theocratic and collectivist New England town that flourished, especially in Massachusetts, Connecticut, and New Hampshire, during the first century of settlement. The town, the basic unit of government and comparable in area to townships in other states, allotted both rural and village parcels to single families by group decision. Contrary to earlier scholarly belief, in all but a few cases settlement was spatially dispersed in the socially cohesive towns, at least until about 1800. The relatively concentrated latter-day villages persist today as amoeba-like entities straggling along converging roads, neither fully rural nor agglomerated in form. The only latter-day settlement experiment of notable magnitude to achieve enduring success was a series of Mormon settlements in the Great Basin region of Utah and adjacent states, with their tightly concentrated farm villages reminiscent of the New England model. Other efforts have been made along ethnic, religious, or political lines, but success has been at best brief and fragile.

Creating the National Domain

With the coming of independence and after complex negotiations, the original 13 states surrendered to the new national government nearly all their claims to the unsettled western lands beyond their boundaries. Some tracts, however, were reserved for disposal to particular groups. Thus, the Western Reserve of northeastern Ohio gave preferential treatment to natives of Connecticut, while the military tracts in Ohio and Indiana were used as bonus payments to veterans of the American Revolution.

A federally administered national domain was created, to which the great bulk of the territory acquired in 1803 in the Louisiana Purchase and later beyond the Mississippi and in 1819 in Florida was consigned. The only major exceptions were the public lands of Texas, which were left within that state's jurisdiction; such earlier French and Spanish land grants as were confirmed, often after tortuous litigation; and some Indian lands. In sharp contrast to the slipshod methods of colonial land survey and disposal, the federal land managers expeditiously surveyed, numbered, and mapped their territory in advance of settlement, beginning with Ohio in the 1780s, then sold or deeded it to settlers under inviting terms at a number of regional land offices.

The design universally followed in the new survey system (except within the French, Spanish, and Indian grants) was a simple, efficient rectangular scheme. Townships were laid out as blocks, each six by six miles in size, oriented with the compass directions. Thirty-six sections, each one square mile, or 640 acres (260 hectares), in size, were designated within each township; and public roads were established along section lines and, where needed, along half-section lines. At irregular intervals, offsets in survey lines and roads were introduced to allow for the Earth's curvature. Individual property lines were coincident with, or parallel to, survey lines, and this pervasive rectangularity generally carried over into the geometry of fields and fences or into the townsites later superimposed upon the basic rural survey. This all-encompassing checkerboard pattern is best appreciated from an airplane window over Iowa or Kansas. There, one sees few streams or other natural features and few diagonal highways or railroads interrupting the overwhelming squareness of the landscape. A systematic rectangular layout, rather less rigorous in form, also appears in much of Texas and in those portions of Maine, western New York and Pennsylvania, and southern Georgia that were settled after the 1780s.

Distribution of Rural Lands

Since its formation, Congress has enacted a series of complex schemes for distribution of the national domain. The most famous of these plans was the Homestead Act of 1862, which offered title to 160 acres to individual settlers, subject only to residence for a certain period of time and to the making of minimal improvements to the land thus acquired. The legal provisions of such acts have varied with time as the nature of farming technology and of the remaining lands have changed, but their general effect has been to perpetuate the Jeffersonian ideal of a republic in which yeoman farmers own and till self-sufficient properties.

The program was successful in providing private owners with relatively choice lands, aside from parcels reserved for schools and various township and municipal uses. More than one-third of the national territory, however, is still owned by federal and state governments, with much of this land in forest and wildlife preserves. A large proportion of this land is in the West and is unsuited for intensive agriculture or grazing because of the roughness, dryness, or salinity of the terrain; much of it is leased out for light grazing or for timber cutting.

Patterns of Farm Life

During the classic period of American rural life, around 1900, the typical American lived or worked on a farm or was economically dependent upon farmers. In contrast to rural life in many other parts of the world, the farm family lived on an isolated farmstead some distance from town and often from farm neighbours; its property averaged less than one-quarter square mile. This farmstead varied in form and content with local tradition and economy. In particular, barn types were localized—for example, the tobacco barns of the South, the great dairy barns of Wisconsin, or the general-purpose forebay barns of southeastern Pennsylvania—as were modes of fencing. In general, however, the farmstead contained dwelling, barn, storage and sheds for small livestock and equipment, a small orchard, and a kitchen garden. A woodlot might be found in the least-accessible or least-fertile part of the farm.

Successions of such farms were connected with one another and with the towns by means of a dense, usually rectangular lattice of roads, largely unimproved at the time. The hamlets, villages, and smaller cities were arrayed at relatively regular intervals, with size and affluence determined in large part by the presence and quality of rail service or status as the county seat. But, among people who have been historically rural, individualistic, and antiurban in bias, many services normally located in urban places might be found in rustic settings. Thus, much retail business was transacted by means of itinerant peddlers, while small shops for the fabrication, distribution, or repair of various items were often located in isolated farmsteads, as were many post offices.

Social activity also tended to be widely dispersed among numerous rural churches, schools, or grange halls; and the climactic event of the year might well be the county fair, political rally, or religious encampment—again on a rural site. Not the least symptomatic sign of the strong tendency toward spatial isolation are the countless family burial plots or community cemeteries so liberally distributed across the countryside.

Regional Small-town Patterns

There has been much regional variation among smaller villages and hamlets, but such phenomena have received relatively little attention from students of American culture or geography. The distinctive New England village, of course, is generally recognized and cherished: it consists of a loose clustering of white frame buildings, including a church (usually Congregationalist or Unitarian), town hall, shops, and stately homes with tall shade trees around the central green, or commons—a grassy expanse that may contain a bandstand and monuments or flowers. Derivative village forms were later carried westward to sections of the northern Midwest.

Less widely known but equally distinctive is the town morphology characteristic of the Midland, or Pennsylvanian, culture area and most fully developed in southeastern and central Pennsylvania and Piedmont Maryland. It differs totally from the New England model in density, building materials, and general appearance. Closely packed, often contiguous buildings—mostly brick, but sometimes stone, frame, or stucco—abut directly on a sidewalk, which is often paved with brick and usually thickly planted with maple, sycamore, or other shade trees. Such towns are characteristically linear in plan, have dwellings intermingled with other types of buildings, have only one or two principal streets, and may radiate outward from a central square lined with commercial and governmental structures.

The most characteristic U.S. small town is the one whose pattern evolved in the Midwest. Its simple scheme is usually based on the grid plan. Functions are rigidly segregated spatially, with the central business district, consisting of closely packed two- or three-story brick buildings, limited exclusively to commercial and administrative activity. The residences, generally set well back within spacious lots, are peripheral in location, as are most rail facilities, factories, and warehouses.

Even the modest urbanization of the small town came late to the South. Most urban functions long were spatially dispersed—almost totally so in the early Chesapeake Bay country or North Carolina—or were performed entirely by the larger plantations dominating the economic life of much of the region. When city and town began to materialize in the 19th and 20th centuries, they tended to follow the Midwestern model in layout.

Although quite limited in geographic area, the characteristic villages of the Mormon and Hispanic-American districts are of considerable interest. The Mormon settlement uncompromisingly followed the ecclesiastically imposed grid plan composed of square blocks, each with perhaps only four very large house lots, and the block surrounded by extremely wide streets. Those villages in New Mexico in which population and culture were derived from Old Mexico were often built according to the standard Latin-American plan. The distinctive feature is a central plaza dominated by a Roman Catholic church and encircled by low stone or adobe buildings.

Reversal of The Classic Rural Dominance

The metropolitanization of life in the United States has not been limited to city, suburb, or exurb; it now involves most of the rural area and population. The result has been the decline of local crafts and regional peculiarities, quite visibly in such items as farm implements, fencing, silos, and housing and in commodities such as clothing or bread. In many ways, the countryside is now economically dependent on the city.

The city dweller is the dominant consumer for products other than those of field, quarry, or lumber mill; and city location tends to determine patterns of rural economy rather than the reverse. During weekends and the vacation seasons, swarms of city folk stream out to second homes in the countryside and to campgrounds, ski runs, beaches, boating areas, or hunting and fishing tracts. For many large rural areas, recreation is the principal source of income and employment; and such areas as northern New England and upstate New York have become playgrounds and sylvan refuges for many urban residents.

The larger cities reach far into the countryside for their vital supplies of water and energy. There is an increasing reliance upon distant coalfields to provide fuel for electrical power plants, and cities have gone far afield in seeking out rural disposal sites for their ever-growing volumes of garbage.

The majority of the rural population now lives within daily commuting range of a sizable city. This enables many farm residents to operate their farms while, at the same time, working part- or full-time at a city job, and it thus helps to prevent the drastic decline in rural population that has occurred in remoter parts of the country. Similarly, many small towns within the shadow of a metropolis, with fewer and fewer farmers to service, have become dormitory satellites, serving residents from nearby cities and suburbs.

MAJOR PROBLEMS OF RURAL SOCIETY IN INDIA

The Rural Society of India is very much backward. It backwardness is very much due to the several problems that haunt the Rural Society. The process of change is very slow and so the problems are more or less age old in recent years, the process of change has been accelerated and so new problems are also cropping up. The major problems consist of the agriculture, the ownership of the land, the lack of cottage industries, lack of education social evils, death of animal, wealth, bad wealth and so on. These problems are the result of traditionalism and conservatism of the Rural Society. The problems of the villages may be studied under the following heads:

1) The problems concerning agriculture; 2) The problems of cottage industries; 3) The problems of population and the family planning animal wealth and animal husbandry; 4) The problems of rural health and education; 5) The problem of the status of women,6) The problem of child marriage, 7) The problem of traditionalism and conservatism 8) The problem of unemployment 9) The problem of land less labour 10) Rural indebtedness 11) Problem of nutrition in villages 12) Problem of housing. Apart from it there are also the problems of untouchability and casteism.

Need for the Study of the Rural Society

1. The study of rural society is essential for carrying out a programme of rural extension
2. It is important to know some of the characteristics of rural society as people have distinguishing features
3. A knowledge of fundamental characteristics of rural situation should contribute much to the understanding of the structure of rural society, the way it functions
4. Rural people are studied in terms of their personal and group relations and as members of groups, organizations and institutions
5. Rural society comprises of all persons residing in administrative unit of village as defined by the authorities

Society is defined as a group of people in more or less permanent association who are organized for their collective activities and who feel that they belong together.

Characteristics of the Rural Society

We have just now seen the students of the Rural Society it is these standards that determinate the characteristics of the Rural Society which may be studied under the following heads:-

1) Too much nearness to nature 2) Agriculture is the main occupation 3) Small size of the society 4) Less density of population 5) Similarity and dissimilarity in the population 6) Similarity and Dissimilarity in the Population 7) Traditional social stratification 8) Social international 9) Social unity Let us take them up one by one and discuss them in detail:-

(i) Too much nearness to nature: We all know that the Rural Society is more or less an unpolluted society. It is nearer to nature and does not suffer from the artificialities that are to be found in the urban society. Rural Society is full of the elements of nature and is also influenced by the philosophy of nature, trees, plants, animals and birds from part of the rural life. In fact all these, things make the rural life very attractive. That is why Rural Society is said to be set in the background of the nature.

(ii) Agriculture being man occupation: In Rural Society people mainly live on agriculture Rural Society and agriculture are more or less synonyms. The entire economic structure of the Rural Society is based on agriculture. Due to agriculture, people have different values and different norms, because agriculture society has a different type of culture.

(iii) Small size of the Society: Urban society is very large and complicated one. As compared to it, Rural Society is quite small in size and simple on structure. Because of this small size inter communication is possible. This changes the whole pattern of the urban way of life.

(iv) Less density of population: In rural areas the population is quite sparse. Because of the lesser number of people in the rural areas, greater relationship between the people is possible. This changes the whole pattern of the social structure in the rural areas.

(v) Similarity and dissimilarity of the population: Rural Society has mutually contradictory characteristics of similarity and dissimilarity in population. Most of the people are agriculture and therefore there is good deal of similarity in the people living in the rural areas. On the other hand people of different castes and also belonging to different cultural patterns live there and because of this there is as good deal of dissimilarity. In spite of this dissimilarity in the groups of people living, there is no complexity in the structure of the society.

(vi) Lack of Social mobility: People living in rural areas are not socially mobile. Generally they do not move form one social status of another and also do not change there occupation. Because of the lack of social mobility the speed of social change in the Rural Society is very slow.

(vii) Traditional Social Stratification: In Rural Society, there is the traditional social stratification, based on the Caste System. Those who belong, while those belonging to the lower caste are considering lower.

(viii) Social inter-action: In Rural Society, the process of social inter-action is not at all organised. In spite of the lack of social inter-action, there is a good deal of nearness amongst the people living in Rural Society. Social activities like co-operation, conflict interference etc. is very much to be found in the Rural Society. It means that on one hand people living in the rural areas are very near to one another and on the other they are also opposed to one another.

(ix) Social Unity: In Rural Society, the element of unity is present in every aspect. The basis of unity is not formal but informal people living in rural areas have common objects, common experiences and sometimes common characteristics, and because of these common factors they live in unity.

RURAL COMMUNITY

Rural Society is like a big family. In fact rural community is nothing but a large family. It does no mean that they have the blood – relationships. In spite of lack of blood relationship there is an element of homogeneity and nearness. They live in co-operation and try to distribute their own resources in the interest of others. This is more to be found in India than on other counties. Rural community in India has the following characteristics:-

(a) Agriculture is the stable Occupation: Most of the people belonging formal community, live on agriculture. They do not use modern means of implements. Apart from it they depend more on fate than anything else.

(b) Lack of density of Population: In rural community there is lack of density of population. There is difference on the basis of religion; caste etc. because of these factors rural community becomes a group of heterogeneous people.

(c) Social Stratification: In rural community, there is the traditional social stratification in the form of Caste System. People belonging to different castes have different types of social statuses. There is division of labour and because of jajmani system every caste is connected with the other.

(d) Co-operation and Conflict: In Indian Rural Community both the characteristics of co-operation and conflict are to be found village people celebrate different festivals together but they also fight on very petty issues of ownership of land, irrigation water etc. the litigation is said to be a very important position of the rural people and represents the elements of conflict.

(e) Ruralisation: It is an important aspect of the Indian Society. It only means that the society is very much concerned with rural characteristics. In rural community this elements of ruralisation is very much present.

LITTLE COMMUNITY

The community which has a small size is known as 'Little Community'. It has all the characteristics of community life, except the size. In most of the villages of India there are little communities. Following characteristics help us to have better idea of the little community.

i) Special position and status ii) Smallness or small size iii) Uniformity and similarity iv) Self dependence and reliance.

(i) Special Position and Status: Little community has a special status in the sense in the sense that is beginning and the end is very clear. Apart from it life in the little community is different from other communities. Its characteristics are sharper and more remarked. Signs of community feeling and collective consciousness are very much to be found there. Because of these characteristics it acquires a special status.

(ii) Smallness or Small Size: Since the little community has a little size, all the members know is another. They have stronger bond of relationship and are helpful to one another. Really speaking little community is a unit in its own. It means that it is more or less complete and self contained. The characteristic of a little community shall be clearer when we discuss another characteristic of self-reliance.

(iii) Uniformity and Similarity: Little community has the characteristic of similarity and uniformity. It means that the activities of most of the members of the little community are smaller. There needs and wants are also more or less similar. They are a classed group and are not prepared to accept changes. Form generation to generation they continue to follow the same pattern and way of life.

(iv) Self –dependence and Reliance: The members of the little community are also to fulfill their needs by themselves. It means that they are a self-dependent self-reliant and self contained group. They do not have any such thing or necessary of want which cannot be fulfilled within that community. As a result of industrialisation and urbanisation the institution of little community is becoming more or less extinct. Their characteristics of self-dependence are also fading away. Many members of the little community go over to towns to earn there love hood and so their characteristics of self –dependence is more or less lost.

GROUP SOCIETY OR FACTION GROUP

Rural Society or the society of the villages has a special feature of faction group or group society. These groups and factions which are mostly based on caste have a special feature. Apart from these characteristics sometimes these groups are formed on the basis of age, family etc. Certain families joint together and form a faction group or group society.

Similarity certain people who belong to the same age group, join together. Apart from it, group is also formed on the basis of common interests. Because of this, it has been seen that sometimes members of the same family are divided into various groups and factors. In fact common interest the desire for security, needs and requirements, feeding a mutual co-operations etc are responsible for the formation of the factions and groups which have special characteristics of the Rural Society.

Sometimes, 'ego' rested interest or instincts of demonstration are also responsible for formation of such groups. Jealousy and ordinary common funds are also sometimes responsible for formation of groups is factions. All these things show that groups and factions are very common in the Rural Society, and they are special features of the Rural Society.

FOLD CULTURE

Folk culture is an important aspect of the human culture. It is a living thing which not only influences the society but also inspires and guides it. It consists of so many things including folk dances, songs, tales proverbs etc. Many of the festivals are nothing but part of our folk culture. India is basically a rural country with predominant of agriculture and so the folk culture of India is predominantly based and influenced by agriculture.

Agriculturists are very conservative are they are not prepared to bring about an accept charges in their way of life and so folk culture of India has not changed as rapidly as it has change in other advanced and industrially developed countries. With industries changing. Influence of western education scientific and technological development and change has also changed scientific and technological development and change has also changed Indian Folk Culture. But the fact remains that a folk culture is the part of village society, and rural life. This is true of India as well.

FOR SOCIETY OR COMMUNITY

That society or community which is very much influenced by the folk life is called folk community or society. It has stronger bound of community feeling. Because of this community feeling there is greater cultural unity. In this context it has to be kept on mind that folk community or society is a part of human society or human life. It has all the characteristics of the human society, or the society in general to which it belongs.

Familism

This special term denotes special importance of the institution of family. In a society in which the collective interests or interest of several persons is given greater importance as compared to the personality of the individual society is said to have the characteristics of familism.

In Indian Society, particularly the village and the Rural Society there is a greater element of familism in Indian village society great stress is laid on the fulfillment of the family obligations and responsibilities and the society also gives greater importance to the collective interests than the interests of the individuals. In Indian Society, particularly the Rural Society, it is the family that forms the basis of political and economic structure. Family responsibility is the most important aspect of man's life. The person is painless to discharge his obligations towards his family; he is not given a place of respect. This is truer of village life then city life. That is why familism is considered to be an importance characteristics and aspect of rural life.

Caste Structure

Caste structure is an important aspect of Rural Society and for proper study of Rural Society. It is necessary that the caste structure should be studied. It is not only an important feature of Rural Society and rural life. Caste is the institution which influences and guides live of the individual member of the society.

Rules and regulations that have framed by the caste are followed by the members of the Rural Society without any bitcher hesitation. Caste system is sometimes religion oriented. But it is not true of every society and of every aspect of life. In one stratification particularly in the village structure of the caste, even today influences the life of the individual.

Dominant Caste

In every village and in every place there is a caste which is most numerous because of its numerical majority or overwhelming number this caste becomes dominant. This dominant caste occupies an important place in the rural life, influences their political and economic activities of that area. In stage and it influences various aspect of life. The Caste System in very strong in the village.

Jajamani System

This is a typical feature of the rural and village society. In fact it is an occupation of a faction. Certain person, who belong to a particular caste fulfill the needs of members of other castes and communities. Same is true of other persons who belong to certain caste who make the member of other castes and communities they are Jajamans is not confined to individual system but to families.

Sometimes member of a particulars family, have their Jajamans throughout the whole village. In other words, it means that all the residents of a village. In other words, it means that all the residents of a village are the Jajamans of a particular family. In fact this Jajaman on economic interest and fulfillment to needs and requirements.

Rural Leadership

Every society has certain members who lead and guide it. Leadership is determined by various factors depending on the nature of the society or the group. In village of Rural Society certainly persons lead it. Normally it is the members of the upper class or the upper caste who are treated as the leaders of the Rural Society or village society. Normally, like other types of societies there are two types of leaders and the Rural Society as well (i) Formal leaders (ii) Informal leaders.

The Basic concepts of Rural Sociology need detailed discussion and attempt to this effect has been made in different chapters it relevant phases here attempt has been made to bring to light the basic concepts of Rural Sociology that form the real subject matter of this growing branch of knowledge which attempts to study the Rural Society in a scientific manner.

CHARACTERISTICS OF INDIAN RURAL SOCIETY

1. Agriculture is main economic activity of rural people. It is based predominantly on Agriculture. Agriculture is the main source of livelihood. The land is distributed between certain families. The distribution of land is between a big land owner and rest of the community, possession of which (land) has prestige value.
2. Caste is dominant institution of village. It is peculiar type of grouping found in rural India. The village is governed to a very great extent by traditional caste occupations, carpenters, cobblers, smiths, washer men, agricultural laborers etc all belonging to separate castes, caste relations are important characteristics of rural life.
3. The religious and caste composition of village largely determines its character. Different castes exist in village due to social distance. The habitation of each caste is separated from others. The habitation area has usually a distinct name e.g. Harijanawada.

4. Each village is independent. All villages have their own organizations, authority and sanctions. Every village has Panchayat which is village self government.
5. Village settlements are governed by certain traditions. The layout of the village, construction of houses, the dress etc is allowed according to the prescribed patterns of the culture of the area. In different areas a certain degree of diversity (differences between villages in the above aspects of the village life) in village organizations is peculiar.
6. The rural society is self-sufficient. The unit of production in rural society is the family, which tries to produce much of its required goods. Economic production is the basic activity of rural aggregates (rural groups)
7. As a territorial, social, economic and religious unit, the village is a separate and distinct entity.
8. It is common to find out a sense of attachment towards own settlement site. In rural society people do not have widely diversified tasks in different parts of the community.
9. Village is characterized by isolation
10. The chief characteristic of rural life is homogeneity, there are not many differences among people pertaining to income, status etc.
11. The other characteristics are less density of population, less social mobility, less education, simplicity, traditionalism, fatalism, believing superstitions etc.

DIFFERENCE BETWEEN RURAL FROM URBAN

First one has to understand that "the Rural" is a Modern concept. Ruralness did not become an issue until some-thing different developed; for example urban or modern. Of course a plur-ality of rural conditions have existed through history, but it didn't become a social/cultural problem until in modern times. Ruralness can be approached from a range of perspectives and can mean many things. Also "rural" conditions and "rural" communities vary; they are different from each other. It is crucial when discussing rural issues to avoid stereo-typing. Rural thus has ambiguous meaning relative to social-historical location of the concept. Besides the geographical nuance that refers to remote or isolated location, various other perceptions can be detected; e.g. social, economic, cultural, political, demographic, etc.

The rapid process of urbanization in modern time is of course a cultural adjustment to economic and social changes taking place. At the turn of the 21st century we have come to sense migration to urban centres as normal and even desirable. (Too) frequently we encounter negative connotations to the "rural" condition. It is often described as inferior to the urban; at best some-thing fragile to be preserved protected, restored or developed.

The field of rural education as opposed to urban education is difficult to define with exact precision for a number of reasons. Too often rural issues are approached in bureaucratic way; e.g. with quantita----tive standards and statistics that do not capture the social character-istics and cultural essence of the communities or human condition dealt with. In this sense rural becomes like a hole. The more material we remove, the larger the hole; the more empty! As numerous people leave a region, the more "rural" it is proclaimed. Of course quantitative data can be important to perceive the rural condition, but probably more qualitative or intuitive insights into the ways of life lived in areas shed clearer light on constrains and trends in the rural condition at the end of the 20th century.

Three types of criteria generally define a rural society: 1) The number of inhabitants, 2) the ratio of inhabitants to open land and 3) type of economic activities – agriculture, fishing, food production and exploitation of raw materials. One problem that one faces is that governments do not use the same statistical criteria for rural and urban populations. For instance, in Japan any cluster of fewer than 30.000 people is considered rural, a definition that would make almost all of the Arctic rural, except a handful of the largest towns.

Nordic countries use various demographic criteria and various concepts to distinguish rural from urban.

Statistics Norway (Statistisk Sentralbyrå) differentiates between densely and sparsely populated areas. Densely populated areas are settlements with more than 200 people and less than 50 metres between the houses. Any agglomeration larger than 200 is registered as an urban settlement. In Northern Norway you find 8 urban settlements with more than 8000 people; the largest is Tromsø with 60000 people.

Another problem defining rural from urban is that the transition from urban to rural is not abrupt, but gradual (the rural-urban fringe). Consequently there are schools recruiting pupils from both urban and rural areas. In order to develop more applicable distinction marks between urban versus rural one can for obvious reasons not apply only demographic criteria. One must consider the type of economic structures and level of technology applied in the settlement, cultural and educational structures, service functions and lifestyle. Settlements depending on natural resources and with a weak industrial structure; may be defined as rural. Absence of industry, medical institutions and research centres are indicators of a rural settlement. And what are the educational opportunities offered within a settlement? Does it offer the opportunity of higher education; college, university college or university.

As consequence of economical and educational structures the composition and demand for labour vary between rural and urban areas Yet another criteria would be to examine lifestyle. Rural societies were typified with their strong adherence to farming and fishing and marked with a high regard for intimacy and traditions. This in contrast with urban areas characterized by impersonal bureaucracy, rationalised specialization and mechanization. Leisure opportunities also differ as urban areas offer their inhabitants a range of cafes and restaurants, theatres and cinemas, galleries, exhibitions, local newspaper etc.

In need of a distinction between rural and urban, which is operational in the research field of rural education, one could consider rural societies as societies characterized by:

1. Relatively few people
2. Depending on natural resources
3. Limited educational opportunities after secondary school
4. Limited job demand and opportunities for skilled and educated labour.
5. Absence of urban commodities like cafes and restaurants, theatres, galleries etc.

The above criteria are helpful in order to separate distinct urban and rural societies. It is however not sufficient for societies in the rural-urban fringe. This for instance applies to societies where we have people living from traditional rural occupations and people commuting to jobs in urban areas. Another example are societies that due to centralization and urbanization processes have experienced growth, with a corresponding expansion in labour opportunities due to expansion and decentralization of some public services and institutions, which in turn has led to a strengthening of local trade and industries. Here we can talk of ongoing urbanization processes in rural societies, were we can study societies which on the one hand have strong links with a typical rural lifestyle, and at the same time enjoy the benefits and disadvantages of urbanization.

Iceland, Greenland and the northern regions of Norway, Sweden and Finland may be characterized as rural regions with urban and rural-urban enclaves.

The cultural aspect of the rural condition is crucial to understand the inter-play of social and personal, natural and cosmological aspect of ruralness. The more people forget the traditional way of life in a community, local myths and sagas that define the essence of the circumstances, the more vague becomes cultural consciousness of the society. It is exactly in this context that teachers and schools have a role to play. It is the teacher's role to teach people understand changes going on in communities and evaluate traditional as well as imported ways of life. To facili-tate understanding of what is important in culture and what is crucial to sustain the communities and natural habitat.

The role of education was often seen as preserving, reproducing, developing the culture of the community. Culture is a way of life; a mentality. As a poli-tical concept in the at the end of the 19th century and the beginning of the 20th, national culture became reconstructed in relation to qualities believed to be attached to tradi-tional rural way of life. This notion of culture can still be detected in poli-tical and educational discourses.

Teacher Education still has a role to play in the development and other changes in the society. Comprehensive research into the changes currently taking place in the rural condition is a fundamental aspect of considering the ways to approach education and schooling in the new rural condition of the 21st century.

However, in the current Nordic context (and presumably the whole Arctic region) the notion of Rural seems to be changing rapidly. Previously the Rural had positive connotation as it referred to cultural preservation; a context were original notions of national culture, language, literature had been saved and cared for at times harsh social change and economic trans-forma-tions, that manifested itself in processes of modernization and urbaniza-tion. In the turmoil of the modern condition references to what was culturally correct tended to be sought in the "simplicity" and "purity" of the Rural context.

This 20th century notion currently appears to be yielding to new vague notion of the Rural. Massive relocation of people towards the Urban centres, migration of immigrants to the country; internationalisation of trade and cultural globalisation seems to secure the new post-modern condition in the Nordic context at the turn of the 21st century. The old notion of rural community is fading away. Agricultural and fishing communities are shrink-ing in terms of population and their economic and cultural relevance is vanishing. In many instances immigrants are hired to keep fishing industries and agriculture alive. In these circumstances the old notion of the rural as culturally pure and nationally original becomes obsolete. At the same time economic importance and cultural relevance of urban centres improve in flow of globalisation; economic rationality and technologisation of the human condition; international pop-culture, mass media, the Internet; genetic technologies, etc.

***

8

RURAL CULTURE

INTRODUCTION

Indian culture is one of the most vibrant cultures in the world and rural culture is the backbone of Indian culture. Agriculture has always been the most important part of rural India, so we have tried to cover the agriculture scenario of rural India with the passage of time. Fairs and festivals are always a reflector of any culture and so is the case with rural India. A glimpse on rural Indian temples and shrines is also given as a separate chapter. Technology is rapidly changing the life styles and we have tried to cover the technological impact on rural India. The urban-rural divide and rural poverty and the possible solution of this problem. It also covers the rural education and the various cultural problems in rural India.

RURAL POVERTY IN DEVELOPING COUNTRIES

The causes of rural poverty are complex and multidimensional. They involve, among other things, culture, climate, gender, markets, and public policy. Likewise, the rural poor are quite diverse both in the problems they face and the possible solutions to these problems. This pamphlet examines how rural poverty develops, what accounts for its persistence, and what specific measures can be taken to eliminate or reduce it.

Broad economic stability, competitive markets, and public investment in physical and social infrastructure are widely recognized as important requirements for achieving sustained economic growth and a reduction in rural poverty. In addition, because the rural poor's links to the economy vary considerably, public policy should focus on issues such as their access to land and credit, education and health care, support services, and entitlements to food through well-designed public works programs and other transfer mechanisms.

About one-fifth of the world's population is afflicted by poverty—these people live on less than $1 a day. Poverty is not only a state of existence but also a process with many dimensions and complexities. Poverty can be persistent (chronic) or transient, but transient poverty, if acute, can trap succeeding generations. The poor adopt all kinds of strategies to mitigate and cope with their poverty.

To understand poverty, it is essential to examine the economic and social context, including institutions of the state, markets, communities, and households. Poverty differences cut across gender, ethnicity, age, location (rural versus urban), and income source. In households, children and women often suffer more than men. In the community, minority ethnic or religious groups suffer more than majority groups, and the rural poor more than the urban poor; among the rural poor, landless wage workers suffer more than small landowners or tenants. These differences among the poor reflect highly complex interactions of cultures, markets, and public policies.

Rural poverty accounts for nearly 63 percent of poverty worldwide, reaching 90 percent in some countries like Bangladesh and between 65 and 90 percent in sub-Saharan Africa. (Exceptions to this pattern are several Latin American countries in which poverty is concentrated in urban areas.) In almost all countries, the conditions—in terms of personal consumption and access to education, health care, potable water and sanitation, housing, transport, and communications—faced by the rural poor are far worse than those faced by the urban poor. Persistently high levels of rural poverty, with or without overall economic growth, have contributed to rapid population growth and migration to urban areas. In fact, much urban poverty is created by the rural poor's efforts to get out of poverty by moving to cities. Distorted government policies, such as penalizing the agriculture sector and neglecting rural (social and physical) infrastructure, have been major contributors to both rural and urban poverty.

The links between poverty, economic growth, and income distribution have been studied quite extensively in recent literature on economic development. Absolute poverty can be alleviated if at least two conditions are met:

- economic growth must occur—or mean income must rise—on a sustained basis; and
- economic growth must be neutral with respect to income distribution or reduce income inequality.

Generally, poverty cannot be reduced if economic growth does not occur. In fact, the persistent poverty of a substantial portion of the population can dampen the prospects for economic growth. Also, the initial distribution of income (and wealth) can greatly affect the prospects for growth and alleviation of mass poverty. Substantial evidence suggests that a highly unequal distribution of income is not conducive to either economic growth or poverty reduction. Experience has shown that if countries put in place incentive structures and complementary investments to ensure that better health and education lead to higher incomes, the poor will benefit doubly through increased current consumption and higher future incomes.

The pattern and stability of economic growth also matter. On the one hand, traditional capital-intensive, import-substituting, and urban-biased growth—induced by government policies on pricing, trade, and public expenditure—has generally not helped alleviate poverty. On the other hand, agricultural growth—where there is a low concentration of land ownership and labor-intensive technologies are used—has almost always helped reduce poverty. Finally, sharp drops in economic growth—resulting from shocks and economic adjustments—may increase the incidence of poverty. Even when growth resumes, the incidence of poverty may not improve if inequality has been worsened by the crisis.

The Rural Poor: Who Are They?

The rural poor depend largely on agriculture, fishing, forestry, and related small-scale industries and services. To understand how poverty affects these individuals and households, and to delineate the policy options for poverty reduction, we first need to know who the rural poor are. The rural poor are not a homogeneous group. One important way to classify the rural poor is according to their access to agricultural land: *cultivators* have access to land as small landowners and tenants, and *noncultivators* are landless, unskilled workers. There is, however, much functional overlap between these groups, reflecting the poverty-mitigating strategies of the poor in response to changes in the economy and society.

Cultivators, who form the bulk of the rural poor in developing countries, are directly engaged in producing and managing crops and livestock. Since these households cannot sustain themselves on the small parcels of land they own or cultivate, they provide labor to others for both farm and nonfarm activities inside and outside their villages. Some members of these households migrate to towns or cities on either a rotational or a long-term basis. In many countries, both small landowners and tenants are under increasing pressure to get out of the agriculture sector altogether. Underlying this process of "depeasantization" are market forces and policies affecting landholdings, rents, prices, credit, inputs, and public investment in social and physical infrastructure.

Noncultivators are perhaps the poorest among the rural poor. Their numbers have been rising rapidly because of the natural increase in population and depeasantization. These workers depend on seasonal demand for labor in agriculture and in rural informal, small-scale industries and services. The landless rural workers are vulnerable to fluctuations in the demand for labor, wage rates, and food prices. They find it even more difficult than small landowners and tenants to gain access to public infrastructure and services. In addition, unlike their counterparts in urban areas, they are often excluded from public sector safety nets (food rations, for example).

Rural women tend to suffer far more than rural men. Their poverty and low social status in most societies is a major contributor to chronic poverty. Substantial evidence from many countries shows that focusing on the needs and empowerment of women is one of the keys to human development.

What Do the Poor Own?

To understand poverty creation in rural areas and its effects on different groups, we need to look at the assets that the poor own or to which they have access, and their links to the economy. The economic conditions faced by the rural poor are affected by a variety of assets (and the returns on them) held at the household, community, and supra-community levels. The poor's *physical assets* include natural capital (private and common property rights in land, pastures, forest, and water), machines and tools and structures, stocks of domestic animals and food, and financial capital (jewelry, insurance, savings, and access to credit).

Their *human assets* are the labor pools—comprising workers of varying ages, genders, skills, and health—in the households and communities. Their*infrastructural assets* are publicly and privately provided transport and communications, access to schools and health centers, storage, potable water, and sanitation. Their *institutional assets* include their legally protected rights and freedoms and the extent of their participation in decision making in households and communities, as well as at the supra-community level. The first two categories of assets are largely regulated through formal and informal networks among individuals and communities. Most rural people, particularly women and those in landless households, are greatly handicapped by inadequate assets and the low and volatile returns on them.

The differences among the rural poor are more clearly reflected in their links to the economy, which determine how they use their assets and participate in production. All of the rural poor are engaged in the production of both tradable and nontradable goods and services. Artisans and unskilled workers provide many nontradable services and some nontradable products (such as staple foods) that small cultivators also produce. Only cultivators, however, have access to small parcels of land through ownership or (sharecropping) tenancy. They are also the only groups of poor people who own or rent physical capital such as tools, implements, and machinery. Artisans and small-scale farmers have only limited amounts of physical capital. They have only limited access to financial capital and acquire it largely through informal agents or institutions, except for tenants, who can use their landlords as conduits to formal credit. Borrowed capital is often costly and is used to maintain consumption during hard times or to buy supplies and equipment needed for farming. Households' labor is used both within the family—for work done by unpaid family members—and to earn the wages paid to landless, unskilled workers in farm and nonfarm activities. All groups of the rural poor are vulnerable to serious risk owing to changes in weather, health, markets, investment, and public policy. The resulting fluctuations in the prices and quantities of their assets and of what they produce can either deepen their poverty or give them opportunities to escape from it. The main reason is that the rural poor have a very low capacity to absorb abrupt financial shocks. In addition, economic crises and natural disasters can bring about sharp increases in poverty and make it more difficult for the poor to escape it.

How Rural Poverty Is Created

Numerous characteristics of a country's economy and society, as well as some external influences, create and perpetuate rural poverty:

- political instability and civil strife;
- systemic discrimination on the basis of gender, race, ethnicity, religion, or caste;
- ill-defined property rights or unfair enforcement of rights to agricultural land and other natural resources;
- high concentration of land ownership and asymmetrical tenancy arrangements;
- corrupt politicians and rent-seeking public bureaucracies;
- economic policies that discriminate against or exclude the rural poor from the development process and accentuate the effects of other poverty-creating processes;
- large and rapidly growing families with high dependency ratios;
- market imperfections owing to high concentration of land and other assets and distortionary public policies; and
- external shocks owing to changes in the state of nature (for example, climatic changes) and conditions in the international economy.

Biases in national economic and social policies can contribute to rural poverty by excluding the rural poor from the benefits of development and accentuating the effects of other poverty-creating processes. Policy biases that generally work against the rural poor include:

- urban bias in public investment for infrastructure and provision of safety nets;

- implicit taxation of agricultural products through so-called support prices and an overvalued exchange rate;
- direct taxation of agricultural exports and import subsidies;
- subsidies for capital-intensive technologies;
- favoring export crops over food crops; and
- bias in favor of large landowners and commercial producers with respect to rights of land ownership and tenancy, publicly provided extension services, and access to (subsidized) credit.

These policies can have both short- and long-term effects on the rural poor. The effects are particularly significant in the context of the structural adjustment programs that many developing countries have undertaken to restore macroeconomic stability and expand the capacity of the economy to increase production, employment, and incomes.

Policies For Reducing Rural Poverty

To design policies that have a chance of effectively helping the rural poor, the focus of policy should be on four major groups:

- small landowners who cultivate their land;
- landless tenants who cultivate other people's land;
- landless laborers who depend on casual or long-term employment in the farm or nonfarm sectors; and
- women, who could also be part of any of the three preceding groups.

All of these groups will benefit from good macroeconomic management—which helps keep inflation in check and maintains unsubsidized prices—because it facilitates sustained economic growth through private investment and competitive markets. Needless to say, unfair laws or poor enforcement of existing laws, exclusion of the poor from decision making, and pervasive corruption in the public sector are no less detrimental to the well-being of the poor than they are to the country's overall economic growth.

Achieving agricultural growth by applying new technologies is one of the most important ways to reduce rural poverty. The impact of such efforts on the rural poor, however, depends on initial conditions, the structure of relevant institutions, and incentives. Research shows that agricultural stagnation has harmed the rural poor in sub-Saharan Africa by creating food shortages and higher prices that have reduced their ability to buy food and find work. Conversely, experience with the Green Revolution showed that rapid agricultural progress made a big difference in reducing rural poverty in parts of South Asia. Researchers have found that higher crop yields reduce both the number of rural poor and the severity of rural poverty. But these effects are strong only if certain conditions are met:

- land and capital markets are not distorted by a high concentration of ownership of natural resources (agricultural land), including unfair tenancy contracts, and repression in the capital markets (with restricted access to finance);
- public policy on pricing, taxes, and the exchange rate does not penalize agriculture and encourage or subsidize labor displacement;
- public investment in basic education and health care is high and used effectively; farmer literacy and good health have great influence on farm productivity;
- public sector support for agricultural research is strong and resulting improvements are made available to small farmers is effective;
- physical capital, like irrigation systems, access roads, is adequately maintained;
- safety nets and social assistance are available for the very poor, particularly the landless (casual) workers and rural women, in the form of public works programs, microfinance, and food subsidies; and
- the rural poor are directly involved in the identification, design, and implementation of programs to ensure effective use of resources and equitable distribution of benefits.

Since the rural poor are a varied group, we need to understand how macroeconomic changes and policies can affect them. The three major ways in which policies affect the rural poor are through markets,infrastructure (including public services), and transfers.

The markets in which the rural poor participate are those for products, inputs (labor and nonlabor), and finance (from formal and informal sources). Several important features of these markets can affect conditions in rural areas.

The infrastructure that directly affects the rural sector's productivity and the rural poor's quality of life includes the economic (transport, communications, extension services, and irrigation) and the social (education, health care, water, and sanitation). Given that most elements of a country's infrastructure are provided through public funding, the level of spending, cost effectiveness, quality of service, and access of the rural poor to infrastructure and public services have important effects on human capital and productivity in rural areas.

Transfers, which are both private and public, provide some insurance against anticipated and unanticipated economic shocks. Most of the rural poor depend on private transfers among households, extended families, and other kinship groups. Public transfers can take the form of redistribution of such assets as land, employment on public works projects, and targeted subsidies for inputs and some consumer products. These transfers supplement or displace private transfers, depending on the policy instrument and how it is used. But these channels—markets, infrastructure, and transfers—do not work in the same way for all of the rural poor because each group has quite different links to the economy.

Key Policy Components Needed to Reduce Rural Poverty

So, what are the key elements when crafting a policy to reduce rural poverty? Competitive markets, macroeconomic stability, and public investment in the physical and social infrastructure are widely recognized as important requirements for sustained economic growth and reduced poverty. In addition, the first requirement of a strategy to reduce rural poverty is to provide the enabling environment and resources for those in the rural sector who are engaged in the agricultural production and distribution system.

Other policy components for national strategies—involving the government, the private (for-profit) sector, and civil society—to reduce rural poverty can include:

Information gathering. The rural poor face many different problems and are not a homogeneous group. Therefore, a sustained effort must be made to gather information about the particular problems they face so that they can be adequately addressed.

Focus on building assets: The government should assess what assets the poor need most to help them earn more. This could be agricultural land or other resources, access to credit, or improvements in health and education. Dependence on raw labor, without a focus on building other assets, is the single most important source of persistent poverty.

The right to adequate land and water: A broad-based land reform program—including land titling, land redistribution, and fair and enforceable tenancy contracts—is critical for reducing rural poverty. It can make small (marginal) landowners and tenants more efficient producers and raise their standards of living.

Basic health care and literacy: The rural poor need to build and strengthen their human capital so they can get out of poverty and contribute more to the economy and society. Basic health care (immunization, provision of clean water, and family planning) and education (literacy, schooling, and technical training)—particularly for women and children—are essential building blocks and should be accessible at reasonable cost.

Local involvement: The infrastructure and services associated with health and education can be funded and maintained best if the target groups are involved in making decisions about the design, implementation, monitoring, and accountability.

Providing infrastructure: The rural poor cannot make the best use of their resources, including human capital, if either the quantity or the quality of some of the key parts of the country's physical infrastructure (irrigation, transport, and communications) and support services (research and extension) is inadequate. The social and physical infrastructure and services can be funded and maintained best—that is, they will be cost-effective and of reasonable quality—if the target groups are involved in designing, implementing, and monitoring them, as well as in ensuring accountability of the government officials responsible for them.

Targeted credit: Informal and formal sources of credit often are too costly for, or unavailable to, the rural poor. Targeted public sector rural credit programs, especially if they are subsidized, benefit the nonpoor far more than the poor. The poor want credit that is available on acceptable terms and when they need it. Recent experiments with community-based credit programs, in which the poor actively participate in the making of lending decisions that are subject to peer accountability, have been successful in reaching target groups at reasonable cost.

Public works: A large and increasing proportion of the rural poor depends on wage labor, because they have either no asset other than raw labor or very few assets: limited quantities of land and domestic animals. A flexible public works program can greatly help the near landless and the landless smooth out household consumption and avoid transient poverty. If it is used on a sustained basis, it can also strengthen the bargaining power of the poor in rural areas.

Decentralized food programs: Some of the rural poor, both individuals and households, suffer from inadequate nutrition most of the time. They need different kinds of support, depending on their circumstances. These may include food supplement programs; food assistance provided through schools, health care clinics, and community centers; and cash transfers. Decentralized and targeted programs seem to work best.

RURAL HERITAGE, RURAL CULTURE AND GLOBAL CULTURE

In the 19th and the beginning of the 20th centuries, the words culture and civilisation became interchangeable. In contemporary society, culture and cultural heritage took on additional meanings, moving to a more holistic approach to these concepts: an approach that takes account of the sociological, political, and aesthetic aspects in parallel to the anthropological and ethnographic ones.

Approaching the Concept of Culture

The term culture receives a variety of interpretations in the literature. Originally the term derives from the Latin cultura stemming from colere, meaning: 'to cultivate' or 'to till the soil' (Hall et al, 2003:7). Thus, culture has been connected to cultivating the mind through learning (the scholar) or cultivating the arts (the artist). Beyond this, the anthropologists, by studying primitive communities, were the first to appreciate that culture is also defined by a set of symbolic meanings that people use to make sense of the world around them, and of certain rules and shared beliefs that enable them to interact with each other and live as a community. Culture, in this sense, includes the values, religious beliefs, norms, social rules, languages and protocols people use to relate to each other, as well as their patterns of behaviour and even the material products they create. This perspective has been transposed to contemporary societies and is today adopted by both ethnographers and sociologists.

The approach to the definition of culture taken by the poet T.S. Elliot in his classical work on this subject (Elliot, 1948) is worth noting. According to Elliot, culture has different associations according to whether we have in mind the development of an individual, a group or the society as a whole. However, these three levels are interdependent - the culture of the individual depends upon the group and the culture of the various groups depend upon the culture of the society which these groups belong to. Therefore it is the culture of society that is fundamental and it is the meaning of culture in relation to the whole society that must be examined first.

It is also worth noting the work of the French sociologist Bourdieu (1984) who studied the participation of people in the part of culture that is marked by aesthetic and literary creation and expression, in other words, the creative arts and literature. The distinction between 'high culture' and 'low culture' is used in association to different population groups, who, according to Bourdieu's research, were found to patronise different forms of art and literature. This distinction has dominated public policy for the arts in the 20th century, with 'high culture taking priority, as being patronised by the best educated parts of society. In contrast, the cultural industries that cater for mass taste, have been often associated with popular culture', i.e. the culture of the majority.

All the above approaches have been put together in a definition provided by Guillaumin (1988:p.41) which is representative of the holistic approach to culture: "the totality of the knowledge and practices both intellectual and material of each of the particular groups of a society, and of a society itself as a whole. From food to dress, from household techniques to industrial techniques, from forms of politeness to mass media, from work rhythm to the learning of family rules, all human practices, all invented and manufactured materials are concerned and constitute in their relationship, culture".

By accepting the above definition, it becomes obvious that there is no normative difference between 'rural' and 'urban' cultures. Culture is a universal concept, which takes meaning from the society it emanates from. Rural communities, therefore, define their own culture in relation to the societies they are part of: the local or regional society, the national society or the global society. Needless to say that an individual as well as a community may participate in more than one societies. An important characteristic of our contemporary societies, due to the great movements of populations through migration, is multiculturalism. Different cultures, being characterised by their own values, beliefs, artefacts, customs etc., are increasingly present in both urban and rural societies across Europe. Culture takes on a different meaning as 'inclusive culture': i.e. accepting the migrant groups' right for cultural expression as well as taking a positive stance towards cross-fertilisation between the cultures of indigenous and migrant communities.

The Value of Rural Cultural Heritage

It is stated by UNESCO that "having at one time referred exclusively to the monumental remains of cultures, heritage as a concept has gradually come to include new categories such as the intangible, ethnographic, industrial heritage as well as natural heritage. The concept of heritage in our time accordingly is an open one, reflecting living culture every bit as much as that of the past." (UNESCO, 2001) Such a radical, dynamic and all-embracing definition of heritage opens new horizons, giving rural heritage the place in cultural policy it deserves and connecting it to all aspects of rural development.

Heritage is manifested through its tangible and intangible products. All aspects that provide us with information about rural people's lifestyles, beliefs, ethics and aesthetics, ways of using their land and of living harmoniously with their surrounding environment, habits and customs, architecture and arts (fine, applied and performing), songs and festivities, rituals and myths, are included.

In the 19th and early 20th centuries, organizations for the preservation of heritage were founded as part of national cultural policy, within Europe and beyond. Museums, theme parks and other heritage institutions became a necessary public amenity, underlining the identity of societies, sustaining the status of the nation to the outside world, and preserving unity on the basis of a shared culture. The open museum, initiated by Hazelius in Sweden in the 19th century as a private nonprofit initiative is a characteristic example of the movement to preserve heritage, and in particular rural heritage. Similar open air museums, which reconstruct villages, village life and the connection of the village to natural resources, sprung all around Europe, celebrating rural heritage. Collections of everyday life objects found their home in countless local museums in European villages; agricultural museums show the means and methods for cultivating the land and eco-museums represent the close links between rural culture, natural environment and the local economy. Countless designations of national and world heritage sites, preservation orders placed in rural settlements of outstanding architectural value, folklore and art festivals organised in most regions, production of traditional handicrafts for markets outside their place of origin, have all given rural her- itage a wide visibility.

Thus, rural heritage was recognised as a source for economic development during the last part of the 20th century. The current interest in rural cultural heritage is also the result of the growing demand by international tourists for places with a special interest, cultural and natural, to enhance their leisure activities. All over Europe, countries are realising, to a greater or lesser extent, the economic benefits of rural tourism searching for possibilities to expand their tourism product through cultural heritage.

Investing in cultural heritage, by restoring it and making it more accessible to the wider public, has already proved very profitable. However, it would be wrong to connect cultural heritage to the development of tourism unconditionally and without considering other development options too.

The question of commercialisation of rural cultural heritage has been the subject of a continuing debate, approached both from a philosophical and an economic development point of view. The involvement of local communities in the decisions about using cultural heritage for local economic development has emerged as a process that might possibly safeguard the interests of the communities themselves, while establishing a sound and balanced basis for exploiting cultural heritage for economic purposes.

A key issue in the debate around rural heritage is the need to understand its meaning and importance for rural communities. The heritage of a place has a different value for different people, according to their purposes. Thus, local communities value their heritage differently from ethnographers or tourists. Also heritage may be attributed a variety of values - symbolic, emotional, historical, aesthetic, scientific etc. Smith (2006) provides a typology of heritage values, as follows:

1. Intrinsic values are based on the views and experiences of cultural heritage by the individual who is an 'insider' of the local culture.
2. Instrumental values are associated with the social or economic aspects of heritage.
3. Institutional values reflect the processes and techniques used by institutions to create public value.

It is thus important to understand how heritage influences the daily lives of people, their social organisation, their economic development and the policies at local and national level. Agenda 21 for Culture provides a framework for turning heritage values to local policies. Based on the UNESCO universal declaration on Cultural Diversity, Agenda 21 also draws clear analogies between cultural and ecological questions.

Rural heritage represents indeed a merging of the natural and cultural elements of rural life, which have always been inseparable features of rural communities. On the natural environment humans have depended for food, shelter, raw materials, and locomotion; moreover, they changed the natural environment through their activities, and in turn the environment dictated their potential for survival. Still, it must be recognized that these two elements, nature and culture, have been studied separately by researchers, who tend to be specialised in natural ecology or culture. The concept of 'cultural environments' links these two elements in a common framework, with significant research and policy implications, stressing the need to view rural territories in a comprehensive way, through the constant interaction of man and environment.

An additional sign of the appreciation of the close links between natural environment and cultural heritage is the analogy between natural ecology and culture which underlies the 'cultural ecosystems' concept, introduced by UNESCO. The anthropologists Julian Steward (1955) and Leslie White (1949) firstly used the term 'ecosystems approach' to describe societies as units interconnected to the whole ecological system.

The term ecosystem was originally used to describe both the biodiversity and the habitat of a place. Changes in one of the ecosystem's parts could immediately generate changes in another (Preucel and Hodder, 1996:24). In parallel to natural ecosystems, cultural ecosystems have been recently defined (UNESCO 2003) as the totality of cultural heritage, past and contemporary, and its relationship to the social and institutional networks that produce or regulate cultural production and the local communities' access to it. The cultural ecosystems have been closely connected with the need for cultural diversity and policies protecting such diversity at local level.

Global Culture in The Rural Context

Global culture is one of the many expressions of globalisation. The term globalisation describes the growing political, social, cultural, economic and technological inter-connectedness and inter-dependence of the world today. It has been argued that as markets and people become more accessible to one another the world is getting smaller. The 'global village' has been coined as early as the late 1940s to describe the effect of electronic mass media in creating a world culture and in collapsing barriers in human communication. As communication technologies advance, cultures continue to overlap and influence each other. In addition, an increase in international trade has created an economic interdependence between many states. This process has effects on the environment, on culture, on political systems, on economic development and prosperity, and on human physical well-being in societies around the world.

Some people see globalisation as a threat to local and national cultures, others see it as an opportunity to transcend borders and benefit from other cultures. To some people it means that everyone is a citizen of the global community and to others it means that nondeveloped countries and regions stand to lose their autonomy, both economic and cultural, as global economic interests take advantage of the formers' vulnerable situation. According to Crane (2003) there are four models for understanding cultural globalisation:

- the cultural imperialism thesis which views cultural globalisation as a kind of cultural domination by powerful nations over weaker nations; and, in a more recent version, the media imperialism thesis which argues that cultural globalisation is the result of the domination over global media channels by international media corporations;

- the cultural flows or network model in which cultural globalisation is conceptualised as occurring through networks that have no clearly defined centres or peripheries;
- the reception theory which examines the responses to cultural globalisation of publics in different countries and regions, and
- a model of national and urban strategies toward cultural globalisation, including the preservation of national and local cultures, resistance to global culture and well as the globalisation of national and local cultures.

Each of these four models is useful for explaining specific aspects of cultural globalisation and the consequences that result from it. The best-known model of cultural globalisation is cultural imperialism, a theory that emerged in the 1960s. This model focuses upon the roles of 'western' governments (mostly USA) and of multinational and transnational corporations in the dissemination of different forms of global culture. It claims that a certain type of mass culture, especially related to the mass media, is diffused from rich and powerful countries that are located at the 'centre' of the world's political, economic and cultural system to poorer and less developed countries on the 'periphery' of the system. This theory, however, has undermined the strength of local and national cultures, the power of networks of cultural agents and the influence of the emerging consciousness of local governments regarding the role of culture in achieving sustainable development. These aspects are expressed by the other models put forward by Crane.

In spite of its weaknesses, cultural imperialism, reconceptualised as media imperialism, remains useful perspective because it can be used to analyse the extent to which some national actors have more impact on global culture than others, and therefore are shaping and reshaping cultural values, identities, and perceptions. Given that the scope and influence of global culture is rapidly expanding, these are important issues. In contrast, the reception model has been used to explain responses to cultural globalisation by publics in different countries. This theory thinks of audiences as responding actively rather than passively to mass-mediated culture and that different national, ethnic, and racial groups interpret the same materials differently. This model does not view globally disseminated culture as a threat to national or local identities. Multiculturalism rather than cultural imperialism is perceived as the dominant trend.

The network model offers an alternative conception of the transmission process, as a set of influences that are not necessarily originating in the same place or flowing in the same direction. Receivers may also be originators. The effect of these cultural flows is likely to be cultural hybridisation rather than homogenisation.

This model is complementary to the fourth approach, which focuses on the strategies used by national or local governments and cultural organisations to respond to cultural globalisation by strengthening and protecting the indigenous cultural basis; and by facilitating the diffusion of local and national cultures to global audiences.

On the basis of this perspective, cultural globalisation is a process that involves competition and negotiation as organisations and audiences attempt to preserve and/or project their cultures in global space. Different countries and regions vary in their emphasis upon preservation of culture as compared to production of culture for global consumption.

Rural people live in the same world as the ones in urban areas. They buy what is available in the global market, they dress according to the global fashion trends, they have mobile telephones, satellite and cable TV, they use the internet. The latter globalises communication by allowing users from around the world to connect with each other via chat rooms, blogs, interactive websites and other www novelties. Rural people and especially the younger generation are not an exception. Access to internet is as much an educational activity as it is a leisure one. Global icons, ideas and cultural products (such as music and films) arrive in rural areas through the internet.

The fact the rural societies are in contact with global culture via the internet does not necessarily mean that they become homogenised or assimilated in a larger and complex global community. We should remember the definition of culture presented in the beginning of this chapter: an individual may belong to several cultural groups and several 'societies' - the local, the national or the global ones. It is not unusual to see young people in rural Greece flock a rock concert one day and go to a traditional fair, singing and dancing under the sounds of traditional music the next day. We live in an era of negotiation, and culture is no different from other aspects of social and economic life. Rural culture has to negotiate with other, non-indigenous cultures in order to survive. This can be more easily achieved if rural culture becomes a feature of integrated development, contributing both to economic gains and sustainability.

CULTURE AS AN AGENT OF CHANGE

Culture is both susceptible to change and resisting it. The dynamic processes which characterise society, i.e. models of interaction between people and the formation of groups or networks, sometimes encourage the acceptance of new ideas while at other times encourage stability (O'Neil, 2006). Three main sources of cultural change can be observed:

1. Internal forces which dictate change, such as technological advancement (invention of new tools, energy sources etc) or cultural evolution or loss (when old cultural patterns are being replaced by new ones).
2. External influences, usually through the contact with other cultures.
3. Changes in the natural environment, which may impose changes for the sake of survival.

O'Neil (2006) provides a model of the different mechanisms of change:

In rural areas, the resistance to change is often quoted as a factor affecting the pace of development, putting the blame on adherence to traditions and established attitudes and practices. At the same time, rural communities are exposed to multiculturalism, not only by being exposed to global communication and global cultural influences but also due to intensive migration from less developed to more developed countries. In this context, culture is increasingly being considered as an important agent of social stability and well being (UNESCO, 2003). Besides being a valuable component of economic development, culture is also seen as a mediator of social inclusion and social cohesion, an agent of change in attitudes and practices that are necessary for achieving tolerance, co-existence and cooperation among the constituent groups of the present-day diverse societies.

RURAL DEVELOPMENT AND CULTURAL CAPITAL

According to Throsby (2003) the definition of capital as goods which, when combined with other inputs (especially labour) gives rise to further goods, has been fundamental to the interpretation of production processes in economics for more that two hundred years. In the context of rural development or local development more generally, different kinds of capital can be identified.

Cultural Capital has received several definitions, which focus on its tangible or intangible characteristics. Tangible cultural capital takes the form of artworks and artefacts, such as heritage buildings, archaeological sites, paintings and sculptures or crafts. Intangible cultural capital comprises artworks which exist as public goods, such as music and literature, as well as the inherited traditions, values, beliefs etc. which constitute the collective consciousness and identity of a group (Throsby, 2003). Cultural capital is both an economic concept, incorporating economic 'goods', and a social concept, based on the consciousness of individuals and communities about their own culture and its normative structures.

Cultural capital is closely linked to social capital. The latter is represented by the nature and extent of personal networks and institutional relationships within a community, created on the basis of trust. It has been suggested that cultural capital may be seen as a 'substance' or a 'quality' of social capital (Jeannotte 2003) while social capital is explained by cultural factors, such as the shared values and understandings that facilitate cooperative behaviour and support collective action. In this way, it is not always easy to separate cultural from social capital, because values, customs and beliefs that influence and shape human behaviour are central to both.

Cultural capital as a component of rural development has therefore several functions. It takes an economic value, which may arise from the number of visitors to a historic building or a festival, the number of tourists attracted to the area due to its cultural assets, the increase of property values, the revenue from artefacts etc.; or a social value, based on collective memory, common identity and shared values within a rural community; or a symbolic value, based on continuity of history, uniqueness, spiritual ties with the built and natural environment, a sense of place, i.e. all those factors that define 'local distinctiveness' and finally an aesthetic or artistic value, which gives the particular cultural product a wider significance, beyond the territory where it is produced.

CULTURAL ENVIRONMENTS

Cultural capital in rural areas incorporates the particular relationship of rural inhabitants with the environment and the natural resources, in all four dimensions described above, i.e. the economic, social, symbolic and aesthetic. Environmental management, landscape formation and landscape aesthetics become part of the cultural capital of a rural community. This special character of rural cultural capital gave rise to the term 'cultural environments' (Downs, 2006). The term has been used from Scandinavia to Australia to embrace or enhance the inseparability of humans and the environment, and has been proposed for wider use as a management and research tool. Within this concept, intangible aspects of cultural heritage can be considered alongside monuments and artefacts, cultural spaces and the environment, and studied as . In this context, environment is no longer seen in solely functional terms as something which determines human behavior or simply as a backdrop for action. Instead, how landscape is shaped through human practice (such as religion, agriculture and animal husbandry, for example)and the qualities of landscape, as defined through experiential approaches are topics that make sense to the communities that live within a landscape as well as to the visitors.

The concept of cultural environments is in harmony with the definition of landscape that is included in the European Landscape Convention: "an area as perceived by people, whose character is the result of the action and interaction of natural and/or human factors" (Council of Europe, 2000, article 1a). Cultural environments constitute a global concern in terms of management, economy, fragile natural and cultural ecosystems and sustainability. How the economic benefits of cultural environments can be maximised and realised through their contribution to sustainable rural development is a subject that needs to be further explored through research and practice - again an important aspect of the European Landscape Convention.

Most environments have been impacted upon and shaped by humans. Such a formation of landscapes through human action impacts the future use of landscapes. This approach puts sustainability very much to the forefront: the past good and bad uses of the landscapes should be considered in thinking about future land use by communities. It is thus important to demonstrate the value of researching past communities and their ecological strategies to present day communities.

How people have sustained and renewed themselves and their environments over time as part of a complex ecological dynamic can be demonstrated by artefacts and material culture, architecture, landscape and the paleo-environment. Also, periods of crisis -varying from natural events such as the effects of major volcanic events, climate deterioration and sea level changes, to changes relating to wider political and economic causes and trends which have catastrophic local and regional effect can be studied. In Scotland, for instance, examples such as the Highland clearances and 'boom and bust' industries such as industrial use of seaweed and the intensive herring industry, have immediate relevance to both the communities and the natural environment. These phenomena inform our understanding of the self-adjusting mechanisms operating in the wider ecological system and therefore inform research into, and policy on, sustainability and renewability.

Traditionally, the 'cultural' and 'natural' aspects of the environment have been managed by separate bodies employing different approaches. The concept of cultural environments links the two aspects together, and puts forward the argument for their common management, to achieve sustainable development. The option of tourism for the development of cultural environments is also considered as the most obvious one, because the combination of authentic built heritage and natural-looking landscapes are commonly the elements of the required tourist experience. The benefits and threats of this option need to be considered carefully, alongside other uses of the land such as agriculture or land development and the need for the protection of both cultural and natural elements of the environment and the identities of the communities that live in them.

This can be achieved only through a wider strategy for the development of methods and procedures to measure, map, monitor and manage cultural environments, and for bringing together these methods and procedures with those which deal with human perceptions of identity and place. Capacity in cultural heritage management and practices must be built and sustained.

COMMUNITY INVOLVEMENT

Since the 1990s it has been widely recognised that a community-based approach to rural development is essential for achieving sustainability. Community-based development planning is important for two reasons: firstly to avoid conflicts and tensions that are likely to occur if local development is not compatible with the community's desires and objectives, especially when outside interests take control of investment and economic activity; and secondly because local pride and confidence is seen as an important precursor to local cohesion, regeneration and sustainable economic development. However, to achieve endogenous development and consolidate the role of local communities as rural development agents, two conditions should be met:

- Firstly, the sustainability objectives become clear to the local community, so that they can control the extent to which their own heritage and culture is threatened by outside interests and excessive commercialisation.
- Secondly, the rural communities do not become introverted by adhering solely to past experiences and heritage, but remain open to new technologies and the challenges posed by globalisation on local and national economies. The modernisation of rural areas and their ability to cope and compete in the global arena does not necessarily imply the loss of their cultural heritage and cultural identity.

A bottom-up approach to rural development has gained support at the level of policy too, as shown by the success of the LEADER Initiative. However, the implementation of LEADER as well as rural regeneration schemes, have revealed that an important condition of success in such an approach is to raise the awareness and the skills of local people to both appreciate and manage their natural and cultural assets. Yet, it should be noted that these very communities may have to deal with substantial social issues such as poverty, unemployment, social exclusion, lack of public facilities and services (e.g. education, health etc). Rural communities need guidance and capacity building to become capable of leading their own future, without jeopardising two important aspects of their development potential: their cultural diversity and local distinctiveness.

***

9

PLANNING RURAL DEVELOPMENT WITH A FOCUS ON CULTURE

CULTURAL DIVERSITY AS AN OBJECTIVE OF LOCAL DEVELOPMENT PLANNING

As mentioned in previous chapters, cultural diversity has emerged as a significant concept at the turn of the 21st century, especially as it related to local development planning. Cultural diversity is a keyword now in the international debates on cultural policy. Also, such terms as cultural development, cultural participation, cultural vitality or cultural democracy are widely used by local governments when planning local development. A commitment to support cultural diversity has gained momentum, following the approval of UNESCO"s Declaration (2001) and Convention (2005) on cultural diversity and the approval of Agenda 21 for Culture (2004). According to the Convention:

"Cultural diversity refers to the manifold ways in which the cultures of groups and societies find expression. These expressions are passed on within and among groups and societies. Cultural diversity is made manifest not only through the varied ways in which the cultural heritage of humanity is expressed, augmented and transmitted through the variety of cultural expressions, but also through diverse modes of artistic creation, production, dissemination, distribution and enjoyment, whatever the means and technologies used".

The above definition is made more specific by the ERIC arts Institute in preparation of the 2007 UNESCO World Report on Cultural Diversity. Three elements are emphasised in making the previous definition more specific: (a) the pluralistic identity and origin of cultural creators, producers, distributors and audiences; (b) the diversity of artistic and other cultural content which diverse audiences can have access to; and (c) the diversity of actors which are involved in decision-making and funding cultural development. It is also emphasised, in the same context, that actions to support and promote cultural diversity achieve sustainability only when a local basis is secured: history, geography, characteristics of the population and vitality of the civil society are the elements, among other things, that define the local basis.

Culture can even be considered as the goal to development at the local level, if development is understood as a possibility to improve the capacities and quality of life of inhabitants, in all the dimensions of human existence (Laaksonen, in UNESCO, 2006). This approach is closely connected to the idea of 'cultural ecosystems' or 'cultural ecology' which allows the implementation of cultural diversity considerations into policies. A cultural ecosystem is encompassing the totality of local cultural characteristics, human capacities for cultural production, local networks that promote culture and the interventions of the cultural industries and local government.

A FRAMEWORK FOR LOCAL CULTURAL POLICIES

The new architecture of effective policies for cultural diversity is based on the potential of cultural ecosystems and incorporates, according to Mercer (in UNESCO, 2006), the following features:

- Forming part of wider policy frameworks and policy objectives, such as rural regeneration schemes, capacity building, retraining and job creation initiatives for the creative industries or civic education, community governance and citizenship development strategies.

- Following an approach to cultural diversity which is inter-departmental, inter-governmental and interdisciplinary, so that resources may be combined and complemented and cultural planning becomes part of an integrated local area planning.
- Understanding the dynamic and cross-fertilising nature of the relationship between public funded culture and the broader commercial and creative industries.
- Accepting the key role of economic development logics as part of a general shift from 'deficit funding' to an 'investment approach' to culture. This new approach has been called, in the context of cultural diversity, 'productive diversity' and has been proved by research in some countries (e.g. Australia) to lead to innovation and capacity building for sustainability in a creative knowledge economy.
- Taking up evaluation of the impacts of cultural diversity policies on local development, through the construction and application of performance indicators. Local governments in several European and American countries have actually started doing this.

The formulation of a cultural strategy is the first step to incorporating culture in rural local development planning. This is however a rather novel concept, which has not yet found wide application, especially in rural areas. A good practice example is provided by the UK. Although there is no national cultural policy in written form, there is an emergent structure for such a policy at local level (including rural areas) building upon the encouragement of every local authority to develop a Local Cultural Strategy. This is not a statutory requirement, but local authorities without a cultural strategy will not be in a strong position to apply, for example, for National Lottery funding to support cultural projects or cultural infrastructure. The existence of a cultural strategy is also a Best Value Performance Indicator which is used in auditing and inspection of the service delivery arrangements of local government by the National Audit Commission and can have significant implications for the funding of cultural services from central government.

DRAWING A CULTURAL STRATEGY TO GUIDE RURAL DEVELOPMENT PLANNING

When drawing a local cultural strategy, it is vital to analyse local resources, to elaborate a common vision of potential development with a focus on culture, and to plan the necessary investments. Three things should be taken into account in the provisions of the strategy:

- The first concerns the influence culture has on the way the local community and local economy is organised: this reflects to a large extent the existence of social capital in the community, the formation is aided greatly by the local cultural capital and the cultural identity of the community.
- The second concerns the community's willingness for experimentation and the ability of rural actors to think in terms of, and undertake projects which integrate culture and heritage in the local economic development process.
- The third concerns the role culture plays in a given territory as an agent of social cohesion and social integration.

The idea of a project is often said to lie at the basis of local development experiments. Projects require an initiative-oriented culture among the local actors (and especially young people) coupled by the will to invest time, effort and capital. Projects allow local agents to communicate with each other, including the public authorities, individuals and local NGOs, to strengthen and even reconstruct the identity of their place, to keep important economic activities in the hands of the local community and attract new (e.g. tourism), to generate positive images for the community, re-forge bonds between people and communities and develop the required complementarities, whether of a commercial or non-market nature (OECD, 2005).

The implementation of projects which build on the cultural heritage and identity of rural communities has been promoted actively by the LEADER Programme with very good results in some cases. The work of PLANED, a community led partnership in the region of Pembrokeshire, West Wales, provides several examples of a project-driven cultural strategy which involves different sections of the local communities in culture based local development initiatives, with emphasis on entrepreneurship.

The success of local initiatives is not, of course, guaranteed. A number of important factors that determine their success should be taken into account, such as:

- Developing a culture-based product that has a value for both local people/producers and visitors/consumers (e.g. tourism product) based on heritage or contemporary culture.
- Making sure that there is community involvement and community control over the implementation of the project.
- Promoting 'new aesthetics' for the consumer /visitor rather than repeating old and worn out models.
- Developing infrastructure and facilities that are us eful for the local community in the long-term, and at the same time serve the visitors/ consumers.
- Sustaining traditional skills that can be used for the preservation of local heritage and for the production of art and craft objects, both traditional and modern.
- Creating employment for local people, especially the young.
- Promoting a culture of entrepreneurship.
- Making sure that appropriate training becomes available to local people, allowing them to build their capacity to take active part in projects.
- Making the area known to a wide audience, working on the ambition to influence the wider culture of the nation or the world.

The culture of a territory has the option to favour or not the social integration of all members of a community, including minorities defined on the basis of origin, ethnicity, language or personal history (e.g. disability, homelessness, addiction etc). A sustainable development strategy should, however, make sure that social integration is promoted and the mutual understanding and cooperation between different social groups is cultivated. Culture can play a major role in this respect, although such a role should not be taken for granted, but actively assisted. Although the Council of Europe, in their 1977 Report painted a rather favourable picture of the effects culture can have on social inclusion, the reality is rather different. Diverse cultural groups often seek to differentiate themselves from others exactly because their cultural signs do not coincide. The example of a project carried out in the island of Santorini, Greece, on the social integration of the large immigrant community of the island through culture, gave us an indication of the necessary components of such a strategy:

- Raising the visibility of the minority culture is the first step to make such a culture known to the local community. Culture facilitates acceptance of others through increased familiarity.
- Promoting networking between the different cultural groups and with those local actors that can influence the community, e.g. local authority or NGOs.
- Building the capacity of the minority cultural groups to express their culture, exercise their creativity and reinforce their identity.
- Creating opportunities for joint production of culture between the minority and majority groups of the community, so that at least part of the local cultural product is shared by all.

IMPLEMENTING A CULTURAL STRATEGY BY SELECTING AND PLANNING DEVELOPMENT ACTIONS

The deployment of cultural activities and cultural products in the implementation of a local development strategy is expected to focus mostly on three types of action: attracting tourism (day visitors or longer-staying tourists); producing creative products; and creating cultural infrastructure. These types of action could be paralleled to sub-sectors of the culture economy of the area. It would be advisable to try and include in a local strategy all three types of action, if possible, in order to diversify the cultural product of the area, address more that one markets and create inter-dependencies and cross-benefits between the sub-sectors of the cultural economy.

Tourism is the single activity that dominates culture-focused development thinking, and rural development planning more generally, because for many constitutes the obvious solution to earn revenue out of the local heritage. Local heritage sites, natural or cultural, museums, monuments, buildings or complexes of architectural or historic interest, attract visitors. If local facilities for accommodation and food are provided, then the visitors may become tourists, and stay longer. This is the principle that has dominated the advent of rural tourism, and many areas have benefited from it. However, rural tourism is a complex economic phenomenon, in a sector that is highly competitive and unstable. A separate local strategy for the development of tourism is necessary, which involves a variety of activities, investment and controls, let alone that it needs to be integrated in a wider regional strategy and a national strategy as well.

The link between culture and creativity also holds prospects for local development. The crafts industry, which is closely related to heritage, is an example that interlinks closely with tourism. Also modern design inspired by heritage has proved a worthy diversification of the art and crafts sector, as vividly shown by the silver jewellery crafts in Greece, Bulgaria, Spain and elsewhere. Some territories famous for a certain type of craft (e.g. porcelain, woodcraft, metal) have invited distinguished modern artists to re-design motifs and forms, so that they appeal to the contemporary consumer. These linkages between arts, culture and creativity have been noted by Clustering Alliance (2004) in some rural communities. The examples they list, all from USA experience, are thought provoking: they range from a plumbing fixture company that has artists-in-residence in rural areas around Wisconsin to design new product lines with an artistic dimension; to a firm in Ohio, which helps local food processing companies design imaginative labelling and tell stories about their products, thus creating brand recognition and making them more attractive and localised. Indeed food and drink are very particular cultural products, which can earn an autonomous place in the local economy, as is the case with appellation d'origin controlee. Media products should also find a place in the local cultural strategy, in a dual function: as publicity material, to aid tourism or the promotion of other cultural products of the area; and as autonomous products, in audio-visual or printed form, recording the heritage (music, dance, photos, heritage sites etc) or contemporary cultural achievements of the local communities and their members.

Creating the right infrastructure for new cultural activities is the most challenging action of all, probably because it involves immediate investment. Infrastructure can range from the creation of a local museum; to building a network of local footpaths; to refurbishing a disused building for cultural use; and many more. The rehabilitation of old buildings, or even former industrial sites or farm complexes, has been given particular attention recently in terms of the cultural uses they may host and the role they may play in the regeneration of declining areas. Although the examples come mostly from urban regeneration sites, rural regeneration may well benefit from such actions, where abandoned or disused buildings and complexes are present. The case of Friche Bell de Mai in Marseille has shown that by offering studio space to artists in the refurbished buildings, the whole area came into life, eventually becoming a creativity centre, fostering artistic collaborations and giving rise to new initiatives, not only in the visual but also in the performing arts.

A local strategy should also introduce planning the necessary actions in stages. At an initial stage, the involvement of the local people should be facilitated through information material, workshops, discussions and campaigns. Often local people do not have the knowledge or the self-confidence to start local development activities based on their culture; and they do not have a picture of the range of cultural activities and cultural products that can be produced/offered in their area, in relation to the capacities within the community and the demand for these activities/products outside the community.

A second step would be to mobilise the local networks and exploit the social capital that is already in existence. The close links between social capital and cultural identity in rural areas makes it imperative to build the cultural strategy on the existing social capital, trying at the same time to increase its range and coverage through the cultural activities themselves. Social capital can both provide solutions for the improvement of business and quality of life in rural communities. The examples from Finland are many, ranging from the building of a local sports centre/community centre through personal labour and donation of materials by villagers; to installing optical lines to increase internet speed and potential for local businesses and homes, enhanced by setting up a village cooperative to manage and fund the process.

A third step would be to plan the cultural activities and goods that will be supported by the cultural strategy and relate them to other aspects of local development, especially employment, quality of life and preservation of local culture and identity. Expert advice will be generally needed for this. The term cultural activities embraces a large variety of actions: restoring a monument, organising a festival, creating a museum, bring out a publication about the area or encouraging the production of cultural goods (art and crafts, digital images, fashion, food etc) are some of the activities.

Planning the activities means that the necessary investment has to be secured, the human resources recruited (from the community or outside it), the right marketing set in place and the inter-connecting facilities and services identified and planned as well. For example, if culture is to become the launching ground for tourism, the complementary facilities of accommodation and catering, and services like transport, must be secured. The local enterprises must be contacted and an audit performed in the area to find out how many of the necessary facilities are already provided, what is the target and how can it be reached through the encouragement of new enterprises, including the incentives that are available for businesses to reach the target.

A fourth step would involve the allocation of responsibilities for the implementation of cultural activities. A share will be certainly taken by the public sector -local government-, another share is expected to be taken by non-profit organisations with a local basis (NGOs) and a considerable share will be taken by local enterprises. A business plan for each activity should then be conducted with a balance sheet of the expected investment and running costs vis-a-vis the expected revenue for at least five years; and a marketing plan for the same period, giving niche markets a good consideration.

A fourth step would be to exploit the advantages of ICT and take part in the 'global village'. This involves an effort to avoid an introverted cultural strategy and open up a window to the world, by offering works of an appealing and meaningful culture which, although having strong local roots, is open to outside influences and takes part in the national and international debate. This would help to make the cultural activities and goods offered locally relevant and accessible to young and old, local and non-local people.

MOBILISING RURAL COMMUNITIES AS CULTURAL AGENTS

The Role of Cultural Agent

We have already talked about mobilising and involving local communities in local development, taking as a starting point the local culture and identity. We will examine in this chapter these qualities and features of rural communities, which, forming part of their culture, can become the springboard for local development.

Rural communities become 'cultural agents' in the sense that they, as culture producers, develop and project these parts of their culture that can be shared with others, providing economic value. Under this framework, a number of issues emerge, which are crucial for understanding the cultural agent's role and the process of activating this role while pursuing local development objectives. These issues examine:

- Rural communities as producers of culture.
- Raising awareness of rural communities' own culture and its value.
- Local identity building through heritage or contemporary cultural production.
- Building on the area's distinctiveness to promote local development.
- Culture as a facilitator of collective action.
- Interaction between the two roles of rural communities, namely those of cultural agents and local development agents
- Effects of multiculturalism on local culture and local development.

To put the role of local communities as cultural agents into context, it is useful to refer to the conceptual model of the Norwegian anthropologist Fredrik Barth (1994) who examined ethnicity in everyday life of people and in the activities of organisations at different levels. According to Barth, we need to refer to three levels of cultural environment in order to understand how local cultural processes develop:

(a) The micro level concerns the everyday life of individuals. At this level cultural entities with clear borders do not exist; on the contrary, individuals differ between them and also one individual can in different situations and at different times assume different identities.

(b) The middle level concerns local communities and their organisation models, socially and administratively. The impetus here is to create a 'sense of community' and motivate people to reach a common goal.

(c) The macro level concerns actors in national and international organisations who create the framework for local activities, through, for example, legislation and government policy. International organisations exercise also influence, through the global economy, the cultural industries or political intervention (e.g. the European Commission).

These levels cannot be separated and individuals may, in their different roles, function at different levels. Thus, interaction between the levels also takes place. According to Barth, at the micro level people formulate their own local identities and culture in constant interaction with their surroundings, while local organisations define local cultures in reciprocity with their wider national and international organisations. In this context, some tension may arise within local communities and between communities and their wider national and global cultural environments, if local and/or national and international organisations produce identities and images that are in conflict with the realities of everyday life in the rural (local) place.

Local Distinctiveness

The term local distinctiveness has been recently used widely to describe what is also referred to as 'sense of place'. Local distinctiveness embodies all that makes a place unique, in terms of nature and culture, history and perceptions. It is a dynamic concept, changing as places change, including the invisible as well as the visible: symbols, festivals and legends, hedgerows, hills and houses (Clifford and King, 2006). Locality is important in defining distinctiveness: archaeology, architecture, landscape, language, food, folklore, events and all things that interact with one another at the level of the street, the neighbourhood, the community. Distinctiveness is not the same as diversity, and involves much more than variety. Just as nature is always experimenting, a locality must also be open to change. However, as change may enrich the locality, it may also deplete its richness. Mass production, fashion, increased mobility, promotion of corporate identity, are just a few factors that have brought to rural areas uniform shop fronts, farm buildings, factories, forests and front doors, while intensive farming has created a bland, unfriendly countryside. In such an environment, it is the detail, the little things that make the difference in the 'sense of place' we have. It is the unusual, the special, the idiosyncratic or the rare that may be important factors in giving a place a sense of itself. But it is also the commonplace –the locally abundant plants, the wall-building materials, the accents and dialects- that defines the context which provides the 'binding force' for the unusual. Unfortunately, much of the fact-gathering, analysis and policy making leave out the very things that make a place distinct, the things that make people love their place and identify with it.

Raising Awareness About Local Distinctiveness

The largest part of the distinctiveness of a rural place stems, therefore, from the cultural heritage of the local communities. All rural areas have a distinct cultural and historic heritage and living 'folk' culture, which is celebrated with traditional folk music, song, dance, and story telling and through the preservation and interpretation of local history and built heritage. These activities enhance community pride, encourage local people to participate in their community, share their knowledge and experience and celebrate their own folk traditions.

However often local people:

- Do not have the knowledge to undertake the technical aspects of researching or safeguarding their heritage.
- Do not have the confidence to perform in public - dancing, storytelling or playing an instrument.
- Can also believe that what is important to them locally to celebrate is not interesting to visitors or tourists.
- Do not appreciate the steps to be taken to interpret, package and market local heritage and culture, so that they become the 'cultural agents' of their locality.

It is very important, therefore, to spend time with local people helping them to develop a more professional approach to cultural activities, encouraging more people to participate. This applies to the activities themselves; to the festivals and events at which local people can showcase these activities. It also pays to try and engage local businesses, helping them realise the benefits of highlighting to potential visitors the range and special nature of local culture and heritage. The example of PLANED provides an illustration of how this can be achieved (Asby, 2007).

PLANED is a local partnership in Pembrokeshire, West Wales, which has taken the form of a non-profit association, helping local communities to become their own 'cultural agents' and 'local community agents'. Local pride and confidence is seen as an important precursor to local cohesion, regeneration and sustainable economic development. Thus, the organisation is assisting local people to value, celebrate and gain economic benefit from their heritage and culture through a number of local projects. One of them is 'Sense of Place', described below in the box.

Celebration is also important to communities, to raise awareness about their own culture and strengthen their identity. This can be done through heritage exhibitions, festivals, competitions and re-enactments of historical events. However, when such celebrations become a tourist attraction, the whole process may become degraded and results, regarding the local identity building, not as expected. An example is provided by the island of Gotland, in Sweden, where the local communities used to organise a 'medieval festival' including re-enactment of an important historical event, dressed up in medieval cloths, singing and dancing traditional songs and eating and drinking traditional specialities.

This festival, which was initiated by local people to celebrate the end of the tourist season for the island and the return to their normal, off-peak occupations, was discovered by tourists and a new wave of tourism was set in motion, based exactly on this festival. However, following this development, the festival lost most of its meaning for local people, since it became a consumptive good for outsiders.

Tourism has been accused to be a threat to local distinctiveness and local cultural identity. Rural places experiencing a high load of tourists risk adapting their culture to the demands of global tourism, at the cost of losing their cultural identity. Global tourists look for something different and original. This may be found in a place where it is possible to create the atmosphere of past times, as seen from the perspective of global tourists. This type of demand may lead to either loss of authentic local culture or 'freezing' it to death.

In the first case, commercialisation may lead to imitations of cultural activities or products, which have been created for mass production, pretending to look like the authentic, but being far removed from it. This has been also discussed in Chapter 4, in relation to cultural goods that are destined for the tourist market.

In the second case, the danger is that local culture will be protected to death, adhering to a 'frozen' past. The image of a 'frozen' culture is contradictory to the varied, heterogeneous and dynamic everyday life of the rural community. Although the image of an authentic culture with roots in former lifestyles is important for many rural inhabitants, especially for identifying with their home region, we have to accept that culture is living and changing, and refusal to adapt continuously is destined to undermine the local culture itself. The image of the past is just one part of the authentic, heterogeneous and changing cultural process that takes place at the local level in rural territories.

Involving the Community in Local Development

Local development is built on partnerships, and one of the basic partners is usually the local authority. Public participation in decisions that concern local development has emerged as a major issue of local democracy in the last 40 or so years. However, involving local communities in decisions about their development is not a simple matter. It requires, firstly, that both community members and administrators have the capacity to enter a dialogue and negotiate productively over possible differences of view and opinion; and secondly, it requires awareness of the different means and methods by which such participation will become effective in local development terms. People can be involved in local development mainly in three ways:

- By taking part in the decisions drawn by politicians and technocrats, through public participation procedures.
- By taking business initiatives themselves, working together as individuals and businesses to improve the economy and quality of their lives.
- By taking collective action, setting up and participating in voluntary organisations of the third sector.

Another important issue in the course of local development is the building of partnerships, the encouragement of collective action and the creation of social capital in a community. Here, we will emphasise some basic points:

- Social capital strengthens the relationships within the local community, making its members more capable to address their collective needs and take action.

- Rural movements promote the participation of civil society in the processes of planning, decision-making and implementation of rural development.
- Local partnerships, bringing together many different actors from the public and private sectors, have been proved to be more effective in terms of local development, than single organisations.

However, to embark on partnerships, it is necessary for a rural community to have established a collective body that would be capable of representing the interests of the community in public decisions. The rural movement in Finland provides an example of the process of building social capital in rural communities through the establishment of a representative collective body, the village association.

Building partnerships at local level, implies that a common vision, a common understanding and common cultural awareness exist among all partners. A method for reaching this awareness is the futures workshops (Jungk & Mullert 1987) used in Finland. In these workshops both local inhabitants and other agents from different regional organisations work jointly to understand and develop further local cultural production. In the workshop, all the participants have an equal opportunity to influence the direction of cultural change in their own locality. The work is always collective, and requires collaboration with different actors from the public, private and voluntary sectors involved in the process. The strength of the method is that it makes possible to reach both individual and organisational views for constructing a common future.

The work of futures workshops is part of the local interactive cultural process, organised to reach a common target. The working process results in common cultural products and cultural images or identities of local communities defined by all the different actors involved in the process. The basic task of futures workshops is to create new ideas for constructing the future of a selected sphere of life in cooperation with other people. For this purpose it has its own special working structure and principles for action. The development work in workshops consists mainly of discussions and the writing of exercises, which are transparent for all participants. The method has the following principles:

- empowerment of each member and cooperation with other people in the workshop;

- work based on the analysis of the present state of affairs of a topic;
- open and rich creation of different ideas concerning the possible futures of the selected topic within a defined time horizon;
- evaluation of the ideas that emerge;
- formulation of a concrete action plan, through which it is possible to reach the desired future from the present situation;
- commitment of participants to realise action plans

THE INTEGRATED SUSTAINABLE RURAL DEVELOPMENT STRATEGY (ISRDS)

The resulting "Integrated and Sustainable Rural Development Strategy" is idealistic in its vision and practical in its focus on mechanisms of implementation. The Integrated and Sustainable Rural Development Strategy (ISRDS) is designed to realize a vision that will "attain socially cohesive an stable rural communities with viable institutions, sustainable economies and universal access to social amenities, able to attract and retain skilled and knowledgable people, who equipped to contribute to growth and development ".The approach outlined in the strategy is applicable and viable for the entire country, and looks toward a horizon of 2010. It will be implemented initially in selected areas of concentration, or pilot areas, and subsequently expanded. The strategy in its totality presents an opportunity for South Africa's rural people to realize their own potential and contribute more fully to their country's future.

The implementation of the ISRDS will use and develop existing institutional, planning, management and funding mechanisms to focus the expenditure of government in the three spheres to more effectively and efficiently respond to needs and opportunities. It is not predicated on additional funding from government. It will increase the efficiency of the application of public funds in rural areas to create appropriate outputs in the places where they are most needed. The structures and procedures of the Medium Term Expenditure Framework will provide the mechanism for realizing national and provincial budget structures to meet ISRDS objectives. Additional funding will be the derived from the structure of delivery relationships that will be established. Sources will include the private sector, public-private partnerships, organs of state and the donor community.

South Africa is in an ideal position to take on board one of the key lessons of the international experience; namely, that successful rural development must be implemented in a participatory and decentralised fashion in order to respond to articulated priorities and observed opportunities at the local level. The reform of municipal government places organs of local government in a central role in integrating programmes to achieve synergistic rural development. Many will need assistance and guidance to develop capacity, but their role and responsibilities are clearly established. They are required to clearly identify local development needs and opportunities and to plan to respond to these. They must align their budgets to achieve their planned objectives. They satisfy the core lessons of international experience as decentralised and accountable entities managing participatory planning and implementation processes. They will be supported by provincial governments which will co-ordinate, integrate and align planning outputs. Provincial governments will also be key agents in the co-ordination and alignment of development inputs–from public and other sources.

At the same time, the Presidency will provide strong strategic leadership and coordination. Each of the line departments will make a critical contribution, and their efforts to achieve synergy and higher effectiveness in their own programmes will be assisted by the strengthened integrative mechanism at the local (municipal) provincial and national spheres.

Towards an Integrated Rural Development Strategy

After the elections of 1999, the government renewed its commitment to rural development, as one of six presidential initiatives announced in the President's address at the opening of Parliament in June 1999.

The intention to initiate an integrated and sustainable rural development strategy was to capitalise on potential synergies among the various government programmes in order to promote and support more rapid and equitable rural development in the country. The presidency initiated a broader process that involved national departments and various stakeholders to formulate the ISRDS and to identify pilot areas for targeted interventions. Since then government departments initiated projects and programmes that target identified areas in the three provinces. Many of these projects and programmes are in the process of being implemented. They were initiated, however, without a clear conception of how the implementation of these programmes would constitute a truly integrated strategy.

The process to formulate the ISRDS is directed by the Deputy President's office, and analytical contributions have been solicited from a number of agencies and organisations. These organisations include amongst others the Development Bank of Southern Africa, the United Nations system, and the World Bank. The Development Bank of Southern Africa submitted their discussion document "Delivering Rural Development" to the Office of the President in July 1999. The World Bank provided a detailed report entitled "Improving opportunities for the rural poor in South Africa: More effective public investment to stimulate rural growth" to the Presidency in June 2000.

Other contributors within the departments and ministries also provided inputs. Also informing the strategy is the Consolidated Municipal Infrastructure Programme (CMIP) document from the Department of Provincial and Local Government, in terms of which the department has identified the need to focus and redirect infrastructure programmes to the rural areas. The primary objective of the programme is to provide funding to municipalities to close gaps in provision of basic levels of infrastructure services to low-income households within a period of ten years and to support the housing programme of government.

Towards Implementing an Integrated Rural Development programme: This document was developed by the Department of Housing to examine the concept of rural areas within the context of the urban rural continuum. It draws on work presented to the Ministers' infrastructure cluster, and work done for the President's Office on the spatial implications of infrastructure as well as proposals by recent mission of the United Nations Food and Agriculture and the proposals by Thuthukani Ma-Afrika on a rural poverty alleviation strategy. Whilst this document was done for a rural housing strategy it also helps to inform the overall Government Integrated and Sustainable Rural Development Strategy.

The Ministry of Minerals and Energy released a document "Towards Implementing an Integrated Rural Development Programme" which articulates the ministerial priorities for the next five years. It covers both mining and energy issues. On mining, the strategic intention is to create integrated and sustainable development in labour and mining areas through the combined and co-ordinated application of the resources and skills of the social partners, being representative of government in all spheres, labour and the mining industry. The results will be measurable reduction in poverty and measurable increases in sustainable job opportunities in affected rural communities.

The strategy will target communities affected by retrenchments with inclusive programmes; target poverty pockets in labour sending and mining areas; cluster delivery into nodes and corridors of sustainable economic development; build on existing rural development work of role players; pursue land rights as a key ingredient for agricultural and other economic development in rural areas and promote the linkages between small-scale mining and tourism. On energy, the objective is to increase access to affordable energy services by rural household by 2010. This is strategic in empowering rural communities.

The International Evolution of Approaches to Rural Development

Worldwide, the pervasiveness of poverty and poor delivery of basic services in rural areas of developing countries continue to constrain development efforts. The prevalence of rural poverty provides major challenges to governments, organisations of civil society and developmental agencies. The failure of many rural development projects during the last three decades has led those involved to consider in more detail the factors that undermined successful outcomes. Prime among these are the issues of inadequate local capacity and the excessive centralisation of decision-making.

As part of a global phenomenon, many developing countries (as well as transition economies and even some western industrialised countries) are now discovering that rural (and urban) communities, if appropriately empowered, can often manage their own local development efforts, and sometimes considerably better than any agency of the state. A properly worked through system of participation and decentralisation holds the promise to provide mechanisms for empowering communities appropriately, though this process is by no means guaranteed.

Decentralisation is often espoused in principle, but undermined in fact by institutional arrangements and financial flows that fail to allow communities to articulate and act on their priorities. While it is true that rural local governments are unlikely initially to be able to take on some of the more sophisticated functions associated with decentralisation, the gains to rural populations—often ignored and effectively disenfranchised by their urban counterparts—are potentially substantial and worthy of closer scrutiny. It is clear that rural areas stand to benefit just as much, and often proportionately more than urban areas from initiatives designed to build the capacity of local governments to manage their own affairs, and empower local communities to take responsibility for their own local development programmes.

As policy-makers increasingly recognise the importance of the interrelationships between political, economic and social reforms, so they are discovering that processes of decentralisation offer a mechanism through which these reforms can be carried out systematically and in parallel. Decentralisation efforts are often accompanied by a broader interest in building more general principles of local empowerment, which include efforts to improve accountability of sub-national governments to their electorates; to increase transparency in their operations; to build more democratic systems of election; to promote greater fiscal autonomy for local authorities; and to build social capital in communities.

Characteristics of Rural Areas

Rural areas throughout the world tend to have similar characteristics. Populations are spatially dispersed. Agriculture is often the dominant, and sometimes the exclusive economic sector, and opportunities for resource mobilisation are limited. These characteristics mean that people living in rural areas face a set of factors that pose major challenges to development. The spatial dispersion of rural populations often increases the cost and difficulty of providing rural goods and services effectively. The specific economic conditions in rural areas result in fewer opportunities than in non-rural locations. Consequently, the tax base is limited, so rural areas are rarely able to mobilise sufficient resources to finance their own development programmes, leaving them dependent on transfers from the centre. Factor markets in rural areas often operate imperfectly, rendering the search for efficient outcomes an extremely challenging one. Furthermore, rural areas are often politically marginalised, leaving little opportunity for the rural poor to influence government policies. In many developing countries, policies have also consistently discriminated against agriculture through high levels of taxation and other macroeconomic policies that have adversely affected agricultural performance and the rural tax base. A net transfer of resources out of rural areas has resulted.

Evolution of Approaches to Rural Development

During the 1970s, in an effort to extend the benefits of development to rural people, a regional or area-based approach was developed and adopted by many countries and by many international agencies. The approach aimed to tackle rural poverty in a cross-sectoral manner through integrated rural development projects (IRDPs). But the mixed experience with IDPs led to the development of a detailed critique, which, among other things, highlighted the failure to involve local people properly in a participatory process and the failure to build capacity as major shortcomings. In addition, an overly centralised, blueprint approach to programmes design left implementation agencies unable to respond to the demands of local people.

The subsequent retreat of national governments and the development agencies from an integrated approach back to more traditional, single-sector approaches left development practitioners looking for new opportunities to support rural development. Recent experience from a number of developing countries suggests that programmes of decentralisation accompanied by parallel efforts to promote greater power and autonomy in decision making for local communities can offer genuine opportunities to improve outcomes.

RURAL AREAS AND RURAL PEOPLE

Rational and effective planning and implementation of rural development will to a large extent depend upon an agreed definition of the rural areas of South Africa. However, there is little agreement on what constitutes rural areas and rural populations, and ambiguities still surround the concept of rural. A related problem is that the rural/ non-urban areas are far from homogenous and include geographical areas and populations that often differ markedly from one another. For example, most concepts of rural areas and populations include commercial farmers, subsistence farmers, some small towns and villages, and nature reserves that have very different – and even contradictory – functions and developmental needs. In South Africa these differences have been complicated by past policies of racial segregation and especially the creation of independent and self-governing homelands that have reinforced some existing divisions and blurred others.

Overlaying these differences are a range of other variations including ecological and natural resources, human settlement patterns, language and cultural differences, lifestyle differences, the proximity or distance of large urban and industrial conglomerates, etc., all of which contribute to the intricacies of the rural picture. Furthermore, because of their poverty and vulnerability, rural households commonly resort to a variety of different strategies to ensure their survival so that it has become more appropriate to describe their economic activities as livelihood strategies rather than jobs or employment.

Framing a successful rural development strategy will therefore depend on taking into account and understanding the diverse and complex realities that constitute the "rural areas". Besides the diversity of the rural areas, successful planning and implementation of rural development will require an understanding of the complex linkages that exist between rural and urban areas.

Within this context, it is necessary to describe the rural population and their problems in greater detail. Some 45% of South Africa's popula tion live in non-urban areas, depending on the definition used. Of these, some 85% live in the former homelands, and the rest on commercial farms and in the small towns. The deepest poverty in South Africa is found in the rural areas. Women form the majority of the rural population and female-headed households are particularly disadvantaged. Three quarters of the children in the rural areas live in households with incomes below the minimum subsistence level. The poorest households also have low levels of literacy and education, difficult and time-consuming access to water, fuel and other services, and few opportunities of gainful employment. This results in high levels of under- and malnutrition, morbidity and mortality of children.

Survey results show that the correlation between poor living conditions, poor access to services, time spent obtaining water and fuel, poor schooling, low incomes, and large household size in rural areas is clear. A basic needs indicator can readily be established according to scores given for the standard of sanitation, water supply, energy supply and housing: On the one hand, 11% of rural households have adequate housing, piped water, and acceptable sanitation; little time has to be spent collecting fuel, and none is taken with water collection. On the other hand, nearly 22% of the rural population lives in abject poverty, with virtually no access to any of these services. These households are largely those in the lowest quintile (20%) of the income spectrum. The nature of poverty across the income spectrum in the rural areas is as follows:

- Households in the three lower quintile income groups spend a great deal of time collecting wood and water. This has important implications for the productivity of family members in income generating activities, and especially for women, who do 90% of the collecting and carrying.
- Educational indicators differ substantially across the five income quintiles, with the less than 20% of adults in the lowest quintile having basic literacy skills, as opposed to more than 50% in the highest quintile.
- While more than a quarter of the poorest quintile has access to agricultural land, their average landholding is less than a third of a hectare (about 3000m2), and 85% of them have communal land.

An understanding of the sources of rural household incomes provides another perspective on the nature of rural poverty: 4.3% of rural households are totally marginalised and have no income; 11.4% are dependent only on pensions; 16.2% are dependent only on unreliable remittances; and 9.9% are dependent on reliable remittances. Women head the majority of these households.

Farm workers are among the poorest and most vulnerable households in the rural areas. Their vulnerability is increased by their dependence on their employers not only for employment and wages, but also for services such as housing, electricity, schooling, access to medical facilities, water and transport. Wages in the agricultural sector fall well below the minimum living level and seasonal workers (who make up a large part of the agricultural labour force) have the added disadvantage of earning approximately 10% less than permanent workers. In addition to their wages, many farm workers receive payment in kind. Furthermore, the educational qualifications of farm workers are low. For example, a survey conducted among black farm workers in 1997 showed that about 50% of farm workers had no schooling and a further 40% had only been educated between grade 2 and 7.

In spite of this very clear concentration of poverty in rural areas, and considerable efforts by government since 1994, very little change in the plight of rural people on the ground has been realised. While various government policy documents have reiterated the importance of targeting state transfers to rural areas, this commitment appears to have lapsed when it came to funding allocations at local levels. A review of the 1998/99 National Budget found that government spending remained skewed in favour of urban areas. For example, while the provincial funds allocation formula is weighted 25% in favour of those provinces with relatively large rural populations, it is clear that this weighted-allocation does not reverse urban bias, as it merely compensates for the higher costs associated with servicing a larger rural constituency. Moreover, research carried out on the allocation and focus of official development assistance received by South Africa shows that only 13% of all these commitments over the past four years have been allocated to rural development. This commitment represents an amount of approximately US$400 million. The ability of the agricultural and rural development sector to access donor funding, however, needs to be compared with its capacity to disburse such funds. The rural sector performed poorly, with donor disbursement rates showing that the sector only managed to spend 9.5 cents of every Rand committed.

KEY ELEMENTS AND VISION OF ISRDS

The vision of the ISRDS is to attain socially cohesive and stable communities with viable institutions, sustainable economies and universal access to social amenities, able to attract skilled and knowledgeable people, equipped to contribute to their own and the nation's growth and development. The strategic intent of the ISRDS is to transform rural South Africa into an economically viable and socially stable and harmonious sector that makes a significant contribution to the nation's GDP. The strategy will benefit the rural poor generally, but particular efforts will be made to target women, youths, and the disabled. A successful strategy to achieve integrated sustainable rural development will reflect each of its three key elements: i.e., integrated, sustainable, and rural development.

Rural Development

Rural development is understood to be multi-dimensional, encompassing improved provision of services, enhanced opportunities for income generation and local economic development, improved physical infrastructure, social cohesion and physical security within rural communities, active representation in local political processes, and effective provision for the vulnerable. Rural development in this context is thus much broader than poverty alleviation through social programs and transfers. The concept places emphasis on facilitating change in rural environments to enable poor people to earn more, invest in themselves and their communities, contribute toward maintenance of the infrastructure key to their livelihoods; in short, to identify opportunities and to act on them. A successful strategy will thus make people less poor, rather than more comfortable in their poverty. This emphasis is complemented by specific measures to assist the vulnerable and relieve the burdens of poverty.

Sustainable

The strategy will be sustainable to the extent that it contributes to increased local growth, that rural people care about its success, and are able to access the resources to keep it going. Thus sustainability implies effective participation to assure that the projects and activities undertaken respond to articulated priorities at the local level. Although initially much of the financing for the strategy may come through line departments and special programs, sustainability over time will increasingly depend on budgets of local governments. A sustainable strategy will thus have to enhance the revenue base of local government over time (although initially the contribution will be small). It will, moreover, have to define a major role for local government in coordinating the participatory decision-making and in mobilizing available resources.

With the upcoming local elections and the creation of the municipal councils in their newly demarcated boundaries, the integrated rural development strategy will provide a framework in which they can become functioning developmental local governments. The institutionalisation of rural development grounded in processes that are the daily work of local governments accountable to their constituents will serve over time to insulate the interventions from becoming overly politicised.

80. Social sustainability is an important dimension of a successful strategy. Rural communities hold a wealth of social capital in the form of extended networks of mutual solidarity, shared beliefs and traditions, and commitments to retain long-standing practices of daily life. Development projects when defined through sound participatory processes, can reinforce and sustain social capital. Conversely, incremental resources brought into rural communities can be divisive and destructive if various groups compete for access through a process that is not generally accepted and understood. One need not idealize the degree of social cohesion within villages. Rural society is not homogeneous, and widespread poverty creates tensions. Growth necessarily brings change, and change can cause conflict. Because the social dynamics of rural areas present challenges to which there are no easy answers, the participatory process should be designed to be as transparent and broadly inclusive as possible, and at least a portion of benefits should be targeted to particular groups that might otherwise be under-recognized, such as women and young people.

Integrated

Integration has been a goal of rural development programmes for many decades. Most of these failed to achieve the desired synergy because they failed to design a mechanism for integration. Rural development is difficult to integrate because it cuts across traditional sectors and involves all levels of government. An effective mechanism for integration will specify what happens at the various levels, who does what, and how the integration will be accomplished. In the South African context, as spelled out in greater detail later, each sphere of government plays a defined role, but the primary locus of integration is at the municipal level through the IDP process. Implementation of decisions arrived at through the IDP process is achieved by drawing on an amalgamated resource envelope comprised of the municipal budget, the commitments of the line departments, and other sources.

A successful strategy incorporating the key elements described above will thus include:

- A vision of the growth process in rural areas;
- A mechanism for integrating existing programs;
- Design for new programmes if needed;
- A defined locus of decision making;
- A meaningful role for local government.
- Clarification of financial flows and channels.
- Key performance indicators or a process for generating them internally to the strategy;
- Procedures to monitor the indicators;
- Sequencing of actions that should take place in the short, medium, and long term.

DYNAMICS OF GROWTH IN RURAL AREAS

There is a dire need to resuscitate the rural economies and advance the cause of rural areas as potential engines of economic growth that would contribute towards their own development and the broad national development agenda. A strategy to achieve growth must be founded on an understanding of how rural areas grow. Although the specific processes vary over time and space, the general dynamics are understood in terms of sources of growth and linkages that spread and multiply the initial impulse. Growth in agriculture, tourism, forestry, and other primary activities that bring incremental earnings into rural areas generates additional incomes through linkage in expenditure and employment.

For example, agricultural growth generates demand for inputs and the retailing activities associated with delivery (transport, financial services, advisory services). Box 2 presented earlier provides a good example of how increased earnings from a tradable agricultural product (in that case wool) generates additional employment and spreads the benefits throughout the community. Farm families spend their higher earnings on goods and services, some of which are produced locally. Moreover, farm families traditionally save a high proportion of income, and a common use of savings is investment in the farm enterprise, thereby contributing to future growth. Linkages with the local economy tend to be higher for small farms than for large farms, in part because they are more labour intensive in their direct operations, and in part because they are likely to purchase labour intensive goods and services (such as outlays for ceremonial occasions, transport, construction from local materials, etc.). For this reason, a land reform program that allows small farmers with low or moderate incomes to purchase entire large farms (in sections) or under-utilised portions of large farms contributes to agricultural growth and stimulates the local economy more than would agricultural growth from the large farm sector.

Empirical studies of expenditure patterns allow estimates of demand for services and infrastructure likely to accompany growth in primary agriculture, forestry, and tourism. For example, expenditure multipliers for transport services tend to be high, as rural people with higher incomes seek increased mobility. Communities will articulate increased demand for roads and bridges, which in turn will lead to a round of private investment in vehicles and increased demand for repair services.

Agriculture need not be the only source of growth, and in many areas it will not be the most important. Natural resources will always be an important determining factor in rural development, as these may be the only resources that some rural areas are endowed with; e.g., tourist sites, water, mineral deposits and others. Where beneficiation of these resources makes economic sense, it will contribute to growth. Any good or service that can be sold outside the community or substitute for something bought from outside will generate a net increment in earnings that can be multiplied.

Rural communities in which agriculture, forestry, tourism, and other activities are growing enter a period of self-reinforcing growth that increases incomes, strengthens transport and communications (thereby reducing the "remoteness" of rural areas), and promotes diversification of the rural economy, as additional incomes are spent on locally produced goods and services. This vision is realistically attainable for rural areas in which the natural resource base supports growth in agriculture, tourism, forestry, and selected other activities.

Much of rural South Africa falls into this category. Other areas have sizeable concentrations of population and limited potential. A realistic assessment must recognize that not all areas have high potential for growth, and that people, particularly the youth, must have opportunities voluntarily to move away from those areas over time. At present most young people in these areas assume that they must move to cities, and many of them will. With a more active land reform, however, growing tourism, and responsive programs of public investment, rural youth in areas of low potential will see new opportunities in more dynamic rural areas and small towns.

Rural growth takes place in a spatial dimension. Rural towns are critical to the developmental opportunities of their hinterlands. Provision of key services in rural towns increases the multiplier for incremental incomes, since rural people can spend more of it closer to home. Much of the infrastructure and service industry (transport, banking, trade, communications, and other) that allows agricultural producers to capture the value of their primary products is located in rural towns. For example, livestock producers come to town to auction their animals and to deliver them to abattoirs. The prices they receive depend directly on the location and quality of these services. Similarly, the service providers will locate only in towns that can provide adequate public services, such as power, communications, public safety, and fire protection. Thus one of the positive outcomes of successful integrated rural development will be a vibrant, rural town that serve their hinterlands and add value to the production that takes place there.

Local economic development can be defined as a functional area within which economic activity, infrastructure, social facilities and human settlement patterns dictate the viability of governmental interventions. The SDI approach has provided valuable experiences on spatial development on a larger scale, which can be used in Local Economic Development (LED). LED in the rural context includes diversifying the economic base, building backward and forward linkages, ensuring basic standards of health and safety through provision of basic infrastructure and services, maximising job creation and building on the inherent potential of local areas. Within the South African ISRDS, local economic development will be stimulated by enhancing synergy of investment and programmes within an area designated as a node. A single node cannot be larger than the regional authority. Several nodes will be clustered to maximise impact wherever possible.

Dynamic growth will benefit many of the rural poor, even those who do not earn the initial incremental incomes. Due to the linkages and multipliers, many poor people will benefit from the second and third rounds of local expenditure, as opportunities for local employment increase and demand for local services rises.

Moreover, as more active rural economies become less cut off, the risks that plague the lives of the rural poor decline. With better roads and more taxis, the chance that a sick or injured child will get to the hospital in time increases for the prosperous and the poor, even though the chances may still not be equal for both.

Rural Safety Net as an Integral Part of The ISRDS

Although growth benefits many of the poor, it clearly does not benefit all. Safety nets are still needed, and the rural poor in most countries are greatly disadvantaged relative to their urban counterparts when it comes to social assistance. Few developing countries include rural people in social security programmes, and the prevalence of self-employment means that unemployment compensation is largely irrelevant to those without work. South Africa is exceptional in this regard, in the sense that many of the key programmes of social assistance extend to rural people and have served since 1994 to prevent much hardship. Government services are by far the greatest services that are both needed and able to reach vulnerable people including those in rural areas. In order to ensure that a safety net is sustained for the vulnerable groups, the efficiency of out-reach capacity of the social service departments co-ordinated by the Department of Social Development will be enhanced. Government should strike a balance between focussing on taking advantage of economic potential and attending to those who are vulnerable and in need of care. The Department of Social Development is presently reviewing the programmes of social assistance, and the findings of that review can be used to identify key changes in the rural safety net that can be introduced to complement the ISRDS.

***

10

Employment in Rural Areas and Rural Market

INTRODUCTION

The context for this concern for clarification of the current employment situation across Rural Europe is also spelled out in the Council Conclusion. In policy terms there are three important "drivers":

(i) The target of a 70% employment rate by 2010, (and a female employment rate of more than a 60%), set by the European Employment Strategy (EES), and the acknowledgement by Commissioner Fischer Boel that rural development policies should help towards meeting these objectives.

(ii) The continuing reform of the Common Agricultural Policy (CAP), including Rural Development Policy. (The shift from product to producer support presents opportunities for rural employment policy which need to be explored and carefully managed, so that their benefits may be maximised.)

(iii) The enlargement of May 2004. (Which has increased very substantially the rural proportion of the EU area/population, and presents rural development questions of a new kind and degree). The wider impacts of these policy drivers are manifest through the special relationship between rural employment and a range of (predominantly public, rather than private) "goods" which are provided by the European countryside. These include, "environmental protection, animal welfare, agricultural product quality improvement, countryside conservation, and protection of the cultural heritage, of biodiversity and of rural tradition and culture…". The sustainability of these is seen as depending upon the "custodianship of the land, which can only be ensured by maintaining a suitable level of employment" (ibid).

The ageing of the rural and farm workforce and the need to accommodate or reduce the flow of young people out of the countryside presents a serious challenge to the sustainability of the European rural economy, and accounts for the emphasis upon demography within the current project. The age structure of the workforce, together with the level of participation by women in many rural regions, also raises important questions in relation to social and economic cohesion.

RURAL LABOUR MARKETS AS SYSTEMS

The report structure above (and indeed the Terms of Reference) is a reflection of an underlying concept of rural labour markets, which is in turn necessitated by the role of the analysis as background and context for an assessment of past (and proposed future) rural policy. It is helpful at this point to make this concept explicit.

A narrow interpretation of the term "labour market" would be associated only with statistical measures of economic activity, employment and unemployment, perhaps including variations in rates between age groups and genders. This report adopts a broader perspective, according to which rural labour markets are viewed as functional systems, in which employment demand is driven by regional competitiveness, whilst employment supply is partly a function of demography, (including inter- regional migration), but also of various socio-economic factors which determine participation rates, and of the qualitative aspects of the labour force often referred to as "human capital". The value of this broader labour market approach is well illustrated with reference to the two-fold role of population migration:

(i) In the short-medium term migration shares with unemployment the role of absorbing imbalances in local supply and demand for labour, and coping with changes in the conditions of job markets. Thus unemployment rates often tell only part of the story, since they may be ameliorated by out-migration, or "inflated" by in-migration over a period of time.

(ii) Since it is selective in its impact on different groups within the population, migration tends to result in second order or "cumulative causation" effects upon regional competitiveness.

DEMOGRAPHY

Demography is probably the single most important supply-side determinant of economic activity and employment trends in rural areas. Two major components need to be considered; (i) Natural increase (the balance between births and deaths) (ii) Migration, - which accounts for the majority of regional change in population, and also has important feed-back effects upon natural increase. The latter are a consequence of selective migration, which results in regional variations in age and gender structures, which are themselves a particular focus of this report.

However, before discussing the two main demographic components it will be helpful to outline rural urban distributions across the EU, and to summarise recent trends. The chapter concludes with a simple typology of European regions according to the nature of recent demographic change.

Broad Historic Trends

In pre-industrial rural society migratory movements were relatively small, and regional population development was predominantly a function of natural population change. Today, with higher mobility, lower fertility rates, and in many cases natural population decrease, regional population development (with regard both to size and structure) is increasingly dominated by inter-regional migratory movements. At more aggregate levels (European or national) the drop in the Total Fertility Rate is the dominant factor behind changes in the age structure (i.e. ageing), though immigration may partly counteract this process.

For much of the twentieth century the regional pattern of population change in most European countries was characterised by a "rural exodus", and increasing urbanisation. Rural populations declined as the young and able left to seek better employment prospects in urban areas. Those remaining formed an increasingly ageing population, with declining local services. This is a well-known story and connected with the advent of industrial society and the urbanisation process.

However, from the 1970s the urban lifestyle has increasingly been rejected in favour of residence in accessible rural areas and commuting. During recent decades the numbers of people migrating away from cities into the nearby countryside or rural areas has increased. "Counterurbanisation" is a common trend in the "well developed" parts of the world, and it has contributed to rising rural populations in many parts of Europe. The rural exodus is now restricted to the remoter and less prosperous regions, and mainly involves young people, who are still attracted to cities as they set out on their careers.

THE RELATIVE IMPORTANCE OF NATURAL CHANGE AND MIGRATION

The relative roles of natural population change and net-migration in total population change between 1995 and 2000, by OECD region type. In the SR and PR regions, net-migration had a dominant and positive effect on total population change. Net-migration reinforces the positive effect of the natural population change in SR regions and hampers the corresponding negative effect in PR. The positive natural population change in PU regions is counteracted by negative changes in net migration but despite this the overall population change is positive. Patterns of natural population development in rural Europe

Basic Concepts and Inter-relationships

Natural population change is simply the difference between number of births and number of deaths during a specified period. Changes in the number of births are a consequence of the development of the birth rates and of the size of the cohorts of childbearing age. Assuming in age-specific fertility rates are stable, large cohorts of childbearing age result in large new cohorts and vice versa. Consequently, the number of births tends to fluctuate as a function of the size of the cohorts in cycles of around 25 years.

From a regional perspective, the process natural population development is inseparably linked to migration, age structure and the size of the cohorts. Thus depopulating and long term out-migration areas tend to have large (residual) elderly cohorts, and therefore low rates of natural increase. At the other extreme metropolitan areas and university towns, where the proportion of persons aged 20-30 years is larger as a consequence of in-migration of people in these ages, tend to have high rates of natural increase.

Key Indicators

Four indicators are useful in the analysis of natural population change:

- Crude birth rate (CBR)
- Crude death rate (CDR)
- Natural population change (CBR-CDR)
- Total Fertility rate (TFR)

Crude birth rates and death rates are expressed per 1,000 head of population. By comparing CBR and CDR (crude birth and death rates) natural population development may be estimated. Although variations in natural population change are predominantly due to differences in age structure, there may also be regional variations in birth rate which are independent of differences in size of age cohorts. These are measured by the total fertility rate (TFR), a theoretical measure, defined as the number of births related to the number of women in the childbearing ages standardised for different cohort sizes. TFR is thus an indicator of differences in fertility which are independent of age structure (which are also included in the crude birth rate (CBR). In the following section each of the four natural change indicators will be discussed in turn.

Patterns of Natural Change

An overview of patterns of natural change across rural and urban regions in the EU. Less than 40% of PR regions showed positive natural change between 1995 and 2000. SR and PU regions were roughly equally split between those which showed natural increase, and those which showed natural decrease (although those which showed increase accounted for a larger share of the population).

Natural population change at the member state level, differentiated by OECD type. The Eastern European countries generally showed negative natural population development during the second half of the 1990s, as did some "old" EU-member states. Both Germany and Italy had a natural population decrease between 1995 and 2000. In Germany this is probably a result of the low fertility rates in combination with ageing in the former DDR and this explanation is probably also valid for Italy.

Of the OECD types, the PR shows the most negative natural change, both when averaged over the EU27, and at the individual member state level. In 11 out of 22 member states in which the comparison is possible the PR regions had the most negative (or smallest positive) average natural change. At the other end of the scale the SR regions had the most positive overall average natural increase, and in 9 member states the SR zone showed the most positive natural increase of the three OECD types. The PU group showed a rather mixed pattern – in eight countries the urban regions had the most positive average change, whilst in four they showed the most negative average.

Patterns of Migration in Rural Europe

In order to provide empirical evidence of the central role played by migration in regional population trends EU25 regional migration patterns for 1995-2000 and 1999-2000 have been estimated. Regression analysis (using the 1999-2000 data) suggests that migration was the prime driver behind regional population change in Europe at this time. The relationship between net-migration and total population development was strong and straightforward; net-migration explains between 83 and 91 percent of the total population development in 1999-2000 (depending on the level of spatial aggregation). The impact of migration on the demographic change is partly direct (in- and outmigration) and partly indirect, (through its impact on the age and gender structure and hence on natural population development).

The Likely Pattern of Future Change in Rural Europe

Sustainability from a Demographic Point of View

Sustainable demographic development implies a balanced process where no sharp breaks occur. Rapid population increases or decreases are considered inconsistent with a sustainable population development.

Other ingredients are that the population structure – e.g. age and gender structure – will not be disturbed in the development process, the reproduction potential is not eroded, that natural population development will not be negative for a long time, and that migratory movements are balanced with respect to different demographic categories. It must, however, be kept in mind that sustainable development is not synonymous with a stagnant population, or a static population structure, neither with increasing population. Instead – even areas with population increase can be in the risk zone as a consequence of low fertility rates, ageing and low reproduction potentials.

Furthermore population decrease can be consistent with sustainable population development – it depends on the effects of the changed population size and structure. Depopulation is not, however, in general consistent with sustainability as the base for a future balanced demographic development is thereby eroded.

RURAL MARKETS

Rural markets in India constitute a wide and untapped market for many products and services which are being marketed for the urban masses. There is a demand for telecommunication services to be provided to in these areas. Till now it was government which was trying to reach the villages through various initiates, but the rural tele-density is very poor and can be improved only through the introduction of modern and suitable technology along with participation from the private operators. The paper here would like to make a strong case for the use of mobile technology for rural areas versus the land line, and that the initiative has to come from the private telecom operators rather than the government end. The various marketing issues related to marketing of telecommunication services in rural areas area seen through the 4 As framework and the experiences of other countries studied for learning. The Indian rural market has a huge demand base and offers great opportunities to marketers. Two-thirds of Indian consumers live in rural areas and almost half of the national income is generated here. The reasons for heading into the rural areas are fairly clear. The urban consumer durable market for products like colour TVs, washing machines, refrigerators and air conditioners is growing annually at between 7 per cent and 10 per cent.

The rural market is zooming ahead at around 25 per cent annually. "The rural market is growing faster than urban India now," says Venugopal Dhoot, chairman of the Rs 989 -crore(Rs billion) Videocon Appliances. "The urban market is a replacement and up gradation market today," adds Samsung's director, marketing, Ravinder Zutshi.

Reasons for Improvement of Business in Rural Area

- Socio-economic changes (lifestyle, habits and tastes, economic status)
- Literacy level (25% before independence – more than 65% in 2001)
- Infrastructure facilities (roads, electricity, media)
- Increase in income
- Increase in expectations

MART, the specialist rural marketing and rural development consultancy has found that 53 per cent of FMCG sales lie in the rural areas, as do 59 per cent of consumer durable sales, said its head Pradeep Kashyap at the seminar. Of two million BSNL mobile connections, 50 per cent went to small towns and villages, of 20 million Rediffmail subscriptions, 60 per cent came from small towns, so did half the transactions on Rediff's shopping site.

Special Features of Rural Market

Unlike urban markets, rural markets are difficult to predict and possess special characteristics. The featured population is predominantly illiterate, have low income, characterized by irregular income, lack of monthly income and flow of income fluctuating with the monsoon winds.

Rural markets face the critical issues of Distribution, Understanding the rural consumer, Communication and Poor infrastructure. The marketer has to strengthen the distribution and pricing strategies. The rural consumer expects value for money and owing to has unsteady and meager status of weekly income; increasing the household income and improving distribution are the viable strategies that have to be adapted to tap the immense potential of the market.

Media reach is a strong reason for the penetration of goods like cosmetics, mobile phones, etc., which are only used by the urban people. Increasing awareness and knowledge on different products and brands accelerate the demand. The rural audience are however critical of glamorous ads on TV, and depend on the opinion leaders who introduce the product by using it and recommending it.

Opinion leaders play a key role in popularizing products and influence in rural market. Nowadays educated youth of rural also influences the rural consumers. Rural consumers are influenced by the life style they watch on television sets. Their less exposure to outside world makes them innocent and fascinated to novelties. The reach of mass television media, especially television has influenced the buying behaviour greatly

Creating Brands for rural India

Rural markets are delicately powerful. Certain adaptations are required to cater to the rural masses; they have unique expectation and warrant changes in all four parameters of product, price, promotion and distribution.

A lot is already emphasized on adapting the product and price in terms of packaging, flavouring, etc and in sachets, priced to suit the economic status of the rural India in sizes like Rs.5 packs and Re.1 packs that are perceived to be of value for money. This is a typical penetration strategy, that promises to convert the first time customers to repeated customers.

The promotion strategies and distribution strategies are of paramount importance. Ad makers have learnt to leverage the benefits of improved infrastructure and media reach. The television airs advertisements to lure rural masses, and they are sure it reaches the target audience, because majority of rural India possesses and is glued to TV sets!

Distributing small and medium sized packets thro poor roads, over long distances, into deep pockets of rural India and getting the stockiest to trust the mobility is a Herculean task. Giving the confidence those advertisements will support. Sales force is being trained to win the confidence of opinion leaders. Opinion leaders play an important role in popularizing the brand. They sometimes play the role of entry barriers for new products.

The method of promotion needs to be tailored to suit the expectations of the market. Techniques that have proved to be successful are Van campaigns, edutainment films, generating word of mouth publicity through opinion leaders, colourful wall paintings. The Wide reach of television has exposed the other wise conservative audience to westernization. Panchayat televisions in Tamilnadu carries message that are well received and contribute to community development.

Dynamics of rural markets differ from other market types, and similarly rural marketing strategies are also significantly different from the marketing strategies aimed at an urban or industrial consumer. This, along with several other related issues, have been subject matter of intense discussions and debate in countries like India and China and focus of even international symposia organized in these countries.

Rural markets and rural marketing involve a number of strategies, which include:

- Client and location specific promotion
- Joint or cooperative promotion
- Bundling of inputs
- Partnership for sustainability

Client and Location specific promotion involves a strategy designed to be suitable to the location and the client. Joint or co-operative promotion strategy involves participation between the marketing agencies and the client. 'Bundling of inputs' denote a marketing strategy, in which several related items are sold to the target client, including arrangements of credit, after-sale service, and so on. Media, both traditional as well as the modern media, is used as a marketing strategy to attract rural customers.

Partnership for sustainability involves laying and building a foundation for continuous and long lasting relationship.

Innovative media can be used to reach the rural customers. Radio and television are the conventional media that are reaching the rural audience effectively. But horse cart, bullock cart and wall writing are the other media, which can carry the message effectively to the rural customers.

Rural marketing is an evolving concept, and as a part of any economy has untapped potential; marketers have realized the opportunity recently. Improvement in infrastructure and reach, promise a bright future for those intending to go rural. Rural consumers are keen on branded goods nowadays, so the market size for products and services seems to have burgeoned. The rural population has shown a trend of wanting to move into a state of gradual urbanization in terms of exposure, habits, lifestyles and lastly, consumption patterns of goods and services. There are dangers on concentrating more on the rural customers. Reducing the product features in order to lower prices is a dangerous game to play.

INDIAN RURAL MARKET

India lives in villages, close to 72 percent of Indian population lives in rural areas. In the country we have 6.36 lakh villages out of which only 13 percent have population above 2000. The rural economy contributes nearly half of the country's GDP (ETIG 2002-03) which is mainly agriculture driven and monsoon dependant. More than 50 percent of the sales FMCG and Durable companies come from the rural areas. The McKinsey report (2007) on the rise on consumer market in India predicts that in twenty years the rural Indian market will be larger than the total consumer markets in countries such as South Korea or Canada today, and almost four times the size of today's urban Indian market and estimated the size of the rural market at $577 Billion.

Census of India defines rural as any habitation with a population density less than 400 per sq. km., where at least 75 percent of the male working population is engaged in agriculture and where there exits no municipality or board, and the same definition being accepted for the paper here. A marketer trying to market his product or service in the rural areas is faced by many challenges; the first is posed by the geographic spread and low population density in the villages in the country. The second challenge is from the low purchasing power and limited disposable incomes in these parts of the country. But this has been changing in the last few decades with agricultural growth rate faster in the 1990's and 80's than the 1970's (CMIE 1996). Green revolution through the introduction of hybrid seeds, fertilizers and systematic irrigation had a major impact on agricultural productivity, and combined with it was a price policy which ensured minimum support price, and in turn insulated the farmers from market risk, cheap input policy and a stable demand (Vyas 2002) . These all lead to a quantum jump in the incomes of farmers in the country. Initially the impact of green revolution could be seen only in the prosperous agricultural states of the country but now slowly its influence has spread across the country with the increase in irrigation (Bhalla & Singh 2001). Though the income levels overall are still very low there are many pockets of prosperity which have come up in the rural areas in the country.

The Features of Indian Rural Markets

1. Large and Scattered Market: - The rural market of India is large and scattered in the sense that it consists of over 63 crore consumers from 5, 70,000 villages spread throughout the country.
2. Major Income from Agriculture: - Nearly 60 % of the rural income is from agriculture. Hence rural prosperity is tied with agricultural prosperity.

   Roughly speaking, a location is defined as "rural", if 75 per cent of the population is engaged in agriculture related activity. In India, close to 70 per cent of the population is agrarian and contributes to about one-third of India's GNP.
3. Low Standard of Living: - The consumers in the village area do have a low standard of living because of low literacy, low per capita income, social backwardness, low savings, etc.
4. Traditional Outlook: - The rural consumer values old customs and tradition. They do not prefer changes.
5. Diverse Socio-Economic Backwardness: - Rural consumers have diverse socio-economic backwardness. This is different in different parts of the country.
6. Infrastructure Facilities: - The Infrastructure Facilities like roads, warehouses, communication system, and financial facilities are inadequate in rural areas. Hence physical distribution becomes costly due to inadequate Infrastructure Facilities.

The rural market is booming beyond everyone's expectation. This has been primarily attributed to a spurt in the purchasing capacity of farmers now enjoying an increasing marketable surplus of farm produce. In addition, an estimated induction of Rs 140 billion in the rural sector through the government's rural development schemes in the Seventh Plan and about Rs 300 billion in the Eighth Plan is also believed to have significantly contributed to the rapid growth in demand. The high incomes combined with low cost of living in the villages have meant more money to spend. And with the market providing them options, tastes are also changing.

Problems in the Booming Rural Marketing

Although the rural market does offer a vast untapped potential, it should also be recognized that it is not that easy to operate in rural market because of several problems. Rural marketing is thus a time consuming affair and requires considerable investments in terms of evolving appropriate strategies with a view to tackle the problems.

Problems in Rural Market

The major problems faced are:

Underdeveloped People and Underdeveloped Markets: The number of people below poverty line has not decreased in any appreciable manner. Thus underdeveloped people and consequently underdeveloped market by and large characterize the rural markets. Vast majorities of the rural people are tradition bound, fatalistic and believe in old customs, traditions, habits, taboos and practices.

Lack of Proper Physical Communication Facilities: Nearly fifty percent of the villages in the country do not have all weather roads. Physical communication of these villages is highly expensive. Even today most villages in the eastern parts of the country are inaccessible during the monsoon.

Media for Rural Communication: Among the mass media at some point of time in the late 502 s and 602 s radio was considered to be a potential medium for communication to the rural people. Another mass media is television and cinemas. Statistics indicate that the rural areas account for hardly 2000 to 3500 mobile theatres, which is far less when compared to the number of villages.

Many Languages and Dialects: The number of languages and dialects vary widely from state to state, region to region and probably from district to district. The messages have to be delivered in the local languages and dialects. Even though the number of recognized languages are only 16, the dialects are estimated to be around 850.

Dispersed Market: Rural areas are scattered and it is next to impossible to ensure the availability of a brand all over the country. Seven Indian states account for 76% of the country's rural retail outlets, the total number of which is placed at around 3.7 million.

Advertising in such a highly heterogeneous market, which is widely spread, is very expensive.

Low Per Capita Income: Even though about 33-35% of gross domestic product is generated in the rural areas it is shared by 74% of the population. Hence the per capita incomes are low compared to the urban areas.

Low Levels of Literacy: The literacy rate is low in rural areas as compared to urban areas. This again leads to problem of communication for promotion purposes. Print medium becomes ineffective and to an extent irrelevant in rural areas since its reach is poor and so is the level of literacy.

Prevalence of spurious brands and seasonal demand: For any branded product there are a multitude of 'local variants', which are cheaper, and, therefore, more desirable to villagers.

Different way of thinking: There is a vast difference in the lifestyles of the people. The kind of choices of brands that an urban customer enjoys is different from the choices available to the rural customer. The rural customer usually has 2 or 3 brands to choose from whereas the urban one has multiple choices. The difference is also in the way of thinking. The rural customer has a fairly simple thinking as compared to the urban counterpart.

The 4A Approach

The rural market may be alluring but it is not without its problems: Low per capita disposable incomes that is half the urban disposable income; large number of daily wage earners, acute dependence on the vagaries of the monsoon; seasonal consumption linked to harvests and festivals and special occasions; poor roads; power problems; and inaccessibility to conventional advertising media.

However, the rural consumer is not unlike his urban counterpart in many ways.

The more daring MNCs are meeting the consequent challenges of availability, affordability, acceptability and awareness (the so-called 4 As).

Availability

The first challenge is to ensure availability of the product or service. India's 627,000 villages are spread over 3.2 million sq km; 700 million Indians may live in rural areas, finding them is not easy.

However, given the poor state of roads, it is an even greater challenge to regularly reach products to the far-flung villages. Any serious marketer must strive to reach at least 13,113 villages with a population of more than 5,000. Marketers must trade off the distribution cost with incremental market penetration. Over the years, India's largest MNC, Hindustan Lever, a subsidiary of Unilever, has built a strong distribution system, which helps its brands reach the interiors of the rural market. To service remote village, stockists use auto-rickshaws, bullock-carts and even boats in the backwaters of Kerela. Coca-Cola, which considers rural India as a future growth driver, has evolved a hub and spoke distribution model to reach the villages. To ensure full loads, the company depot supplies, twice a week, large distributors which who act as hubs. These distributors appoint and supply, once a week, smaller distributors in adjoining areas. LG Electronics defines all cities and towns other than the seven metros cities as rural and semi-urban market. To tap these unexplored country markets, LG has set up 45 area offices and 59 rural/remote area offices.

Study on buying behaviour of rural consumer indicates that the rural retailers influences 35% of purchase occasions. Therefore sheer product availability can affect decision of brand choice, volumes and market share. Some of the FMCG giants like HLL took out project streamline to significantly enhance the control on the rural supply chain through a network of rural sub-stockists, who are based in the villages only. Apart from this to acquire further edge in distribution HLL started Project Shakti in partnership with Self Help groups of rural women.

Affordability

The second challenge is to ensure affordability of the product or service. With low disposable incomes, products need to be affordable to the rural consumer, most of whom are on daily wages. Some companies have addressed the affordability problem by introducing small unit packs. Godrej recently introduced three brands of Cinthol, Fair Glow and Godrej in 50-gm packs, priced at Rs 4-5 meant specifically for Madhya Pradesh, Bihar and Uttar Pradesh – the so-called 'Bimaru' States.

Hindustan Lever, among the first MNCs to realise the potential of India's rural market, has launched a variant of its largest selling soap brand, Lifebuoy at Rs 2 for 50 gm. The move is mainly targeted at the rural market. Coca-Cola has addressed the affordability issue by introducing the returnable 200-ml glass bottle priced at Rs 5. The initiative has paid off: Eighty per cent of new drinkers now come from the rural markets. Coca-Cola has also introduced Sunfill, a powdered soft-drink concentrate. The instant and ready-to-mix Sunfill is available in a single-serve sachet of 25 gm priced at Rs 2 and multiserve sachet of 200 gm priced at Rs 15.

Acceptability

The third challenge is to gain acceptability for the product or service. Therefore, there is a need to offer products that suit the rural market. One company, which has reaped rich dividends by doing so, is LG Electronics. In 1998, it developed a customised TV for the rural market and christened it Sampoorna. It was a runway hit selling 100,000 sets in the very first year. Because of the lack of electricity and refrigerators in the rural areas, Coca-Cola provides low-cost ice-boxes – a tin box for new outlets and thermocol box for seasonal outlets.

The insurance companies that have tailor-made products for the rural market have performed well. HDFC Standard LIFE topped private insurers by selling policies worth Rs 3.5 crore in total premia. The company tied up with non-governmental organisations and offered reasonably priced policies in the nature of group insurance covers.

Awareness

Mass media is able to reach only to 57% of the rural population. Creating awareness then, means utilizing targeted, unconventional media including ambient media .For generating awareness, events like fairs and festivals, Haats, etc., are used as occasions for brand communication. Cinema vans, shop-fronts, walls and wells are other media vehicles that have been utilized to increase brand and pack visibility. Ideas like putting stickers on the hand pumps, walls of the wells putting on tin plates on al the tree surrounding the pond are some of the innovative media used by personal wash like Lux and Lifebuoy and fabric wash items like Rin and Wheel. Idea was to advertise not only at the point of purchase but also at the time of consumption.

With large parts of rural India inaccessible to conventional advertising media – only 41 per cent rural households have access to

TV – building awareness is another challenge. Fortunately, however, the rural consumer has the same likes as the urban consumer – movies and music – and for both the urban and rural consumer, the family is the key unit of identity. However, the rural consumer expressions differ from his urban counterpart. Outing for the former is confined to local fairs and festivals and TV viewing is confined to the state-owned Doordarshan. Consumption of branded products is treated as a special treat or indulgence.

Hindustan Lever relies heavily on its own company-organised media. These are promotional events organised by stockists. Godrej Consumer Products, which is trying to push its soap brands into the interior areas, uses radio to reach the local people in their language.

Coca-Cola uses a combination of TV, cinema and radio to reach 53.6 per cent of rural households. It doubled its spend on advertising on Doordarshan, which alone reached 41 per cent of rural households. It has also used banners, posters and tapped all the local forms of entertainment. Since price is a key issue in the rural areas, Coca-Cola advertising stressed its 'magical' price point of Rs 5 per bottle in all media.LG Electronics uses vans and road shows to reach rural customers. The company uses local language advertising. Philips India uses wall writing and radio advertising to drive its growth in rural areas.

The key dilemma for MNCs eager to tap the large and fast-growing rural market is whether they can do so without hurting the company's profit margins. In case of nestle, company's product portfolio is essentially designed for urban consumers which cautions companies from plunging headlong into the rural market as capturing rural consumers can be expensive.

Role of Rural Retailing

Retailing is the final phase of the distribution channel and it is clear by now that it is availability and distribution that drives growth in rural Indian markets. Hence retailing will be significant and will undergo greater organisation and maturity as is being witnessed in the urban markets, even in the rural markets. Innovative retail models which take into account the nuances of rural markets is the way forward.

Study on buying behaviour of rural consumer indicates that the rural retailers influences 35% of purchase decisions. Therefore sheer product availability can affect decision of brand choice, volumes and market share. India offers a huge, sustainable and growing rural market which can be tapped effectively through innovative distribution channels with retailing being the most critical element of this strategy as it is the final touch point and the actual touch point with the customer which can be the most critical influence in the buying process.

RURAL MARKETING IN INDIA ECONOMY

The concept of Rural Marketing in India Economy has always played an influential role in the lives of people. In India, leaving out a few metropolitan cities, all the districts and industrial townships are connected with rural markets.

The rural market in India is not a separate entity in itself and it is highly influenced by the sociological and behavioral factors operating in the country. The rural population in India accounts for around 627 million, which is exactly 74.3 percent of the total population.

The Registrars of Companies in different states chiefly manage: The rural market in India brings in bigger revenues in the country, as the rural regions comprise of the maximum consumers in this country. The rural market in Indian economy generates almost more than half of the country's income. Rural marketing in Indian economy can be classified under two broad categories. These are:

- The market for consumer goods that comprise of both durable and non-durable goods
- The market for agricultural inputs that include fertilizers, pesticides, seeds, and so on

The concept of rural marketing in India is often been found to form ambiguity in the minds of people who think rural marketing is all about agricultural marketing. However, rural marketing determines the carrying out of business activities bringing in the flow of goods from urban sectors to the rural regions of the country as well as the marketing of various products manufactured by the non-agricultural workers from rural to urban areas. To be precise, Rural Marketing in India Economy covers two broad sections, namely:

- Selling of agricultural items in the urban areas
- Selling of manufactured products in the rural regions

Some of the important features or characteristics of Rural Marketing in India Economy are being listed below:

- With the initiation of various rural development programmes there have been an upsurge of employment opportunities for the rural poor. One of the biggest cause behind the steady growth of rural market is that it is not exploited and also yet to be explored.
- The rural market in India is vast and scattered and offers a plethora of opportunities in comparison to the urban sector. It covers the maximum population and regions and thereby, the maximum number of consumers.
- The social status of the rural regions is precarious as the income level and literacy is extremely low along with the range of traditional values and superstitious beliefs that have always been a major impediment in the progression of this sector.
- The steps taken by the Government of India to initiate proper irrigation, infrastructural developments, prevention of flood, grants for fertilizers, and various schemes to cut down the poverty line have improved the condition of the rural masses.

***

11

Lifestyle and Health Status in Rural Society

INTRODUCTION

Compared with people living in more populated urban areas, residents in rural environments may experience greater isolation, have less access to health care, and identify with more traditional cultural beliefs about health and illness. Rural residents often seek care from a health professional who may also be an extended family member, neighbour, friend or professional colleague. The rural health care provider should consider these factors, along with other social, economic, and cultural contextual attributes, when he or she responds to ethical situations involving his or her patients' care. The ethical situations include, among others, the often overlapping professional and personal roles of rural health care providers, and the threats to patient confidentiality and privacy that may occur in smaller communities. The rural context has factors that can cause or impact such situations, including geographic isolation, small population, and close social and/or kinship relationships among community members. These factors create unique ethical opportunities and challenges for rural health care providers. Clinicians and administrators should seek potential ethics resources, such as ethics programs and consultants, and the ethics literature, to help solve conflicts. Despite the uniqueness of every rural community, there are several general characteristics that influence ethics conflicts in all rural communities. There are limited resources in the rural setting, including various health care specialties as well as experts in the field of ethics.

Clinicians often work in isolated situations, away from peer colleagues. There are geographic barriers such as challenging roads, limited public transportation, and weather conditions that may impede patients' access to health care services. Rural clinicians commonly experience the phenomenon of overlapping or dual relationships with patients. Because everyone knows each other, disease stigma is frequently encountered, especially associated with conditions like substance abuse, mental illness or sexually transmitted diseases. Professional isolation and high demands for the clinician's services can create provider stress. In some cases a doctor or nurse is on call 24x7 with no real backup. These general contextual features regularly surround and shape the ethical issues that are encountered by rural clinicians. When ethical issues occur in small towns, they are often not easily remedied using professional guidelines that were, in many cases, developed and applied in urban-based and resource-rich health care facilities. The rural clinician's response to ethics conflict is based not only on his or her training, experience, and professional guidelines, but also on the community's values. The social, economic, and cultural characteristics of small communities reflect the values of their residents.

Understanding the health care ethics conflicts that have occurred in the past in the rural community can provide insights to clinicians, enabling them to more effectively respond to ethics conflicts and anticipate potential conflicts.

DEFINING RURAL COMMUNITIES

The term "rural" can be defined in many ways, depending both on who is defining it, and why he or she is defining it, which can result in sometimes confusing and conflicting statistics. In fact, the National Rural Health Association does not provide a definition of rural, but instead encourages individuals to tailor the term to meet the needs of a particular program. Population is a common way to define rurality. According to the U.S. Bureau of Census, about 60 million people live in areas defined as "rural." Rural residents make up about one-fifth (20%) of the total U.S. population

and are spread across four-fifths (80%) of the U.S. land area. Another common definition of rural is based on the geographic size of a community relative to population density; for example, the number of people living in a square mile. More remote regions having fewer than six people per square mile are defined as frontier areas, although this statistic varies as well, depending on the program. Of the total U.S. rural population, about 5% live in towns of 2,500 residents or less. Some definitions of rural also consider the distance to services and/or "time to access services" (e.g., greater than 30 minutes or more than 20 miles to a certain destination). It is important to note that the time to access health care may also impact residents who live in inner city or suburban areas due to transportation challenges. Perceptions of "rural" and the available resources within a particular region are relative in nature. For example, some communities with a population of approximately 25,000 may statistically be defined as rural, yet have features that one expects to find in a large city. Or, in a relative sense, residents in a town of fewer than 2,500 may perceive a community with a population of 10,000 as a city. Likewise, a family living in a frontier region may not feel isolated, because urban-based services are relatively easy for them to access via telecommunication and reliable transportation.

RURAL DEMOGRAPHICS

American rural communities have a diversity of geographical regions, population dynamics, age stratification, and other features. For instance, a higher proportion of African-Americans reside in the southeast, more elderly live in the rural Midwest, while Native Americans tend to reside on or near reservations (usually located in more remote parts of the United States). Rural communities also have diverse population trends; some have experienced a declining population in recent years, and others have had an economic revival and population growth.

Demographically, the population in rural communities sometimes is described as bipolar; rural residents tend to be under the age of 17, or over 65 years of age. This may be because many rural communities have strong schools and are safe places to retire, but lack the employment options needed to sustain groups of young and middle-aged adults. Consistent with national trends, there has been an increase in racial and ethnic minorities in rural areas, who now comprise about 17% of the overall rural population. Overall, compared to urban residents, a greater proportion of rural residents are married, have fewer years of formal education, have lower incomes, and fewer have completed high school.

RURAL CHARACTERISTICS AND LIFESTYLE

Even though each rural community is unique, the overall experience of living in a small rural town It's always risky to generalize about a particular group, as this can perpetuate stereotypes. Furthermore, the belief systems of rural residents are complex, and there is wide diversity among and within communities. However, some authors describe a "rural culture" associated with local belief systems and values, including some rural residents' sense of fatalism and subjugation to nature associated with nature-oriented businesses (e.g., agriculture, mining, timber production). Weather also tends to play a prominent role in these residents' lives, because of its effect on economic activities and the ability to travel.

Conservative Values And Perspectives

Members of small and homogenous communities tend to be conservative politically and socially, with some exceptions. They tend to be "church-going," i.e., actively involved in a faith community's activities. Residents in rural communities are likely to adhere to traditional gender-role expectations, holding precise ideas about what constitutes "men's work" versus "women's work." Men are expected to work outside the home for money and to support their families without public assistance. Women are expected to manage the household and to nurture and support their husbands and children without monetary compensation. Such old-fashioned, or (what some might call) sexist expectations shape some rural people's views about medical providers. Regardless of their actual education, women tend to be viewed as the "nurses" while men are considered "doctors." Female clinicians living in a rural community may be expected to serve as a community resource, and volunteer their professional expertise. Community members may look to nurses and other clinicians to provide no-cost consultations, to respond to local emergencies, or even to provide home-care services that are otherwise unavailable.

Characteristics of the Rural Lifestyle and Culture

- Extensive distances between people and services
- Work and recreational activities are often cyclic and seasonal in nature
- Prominence of high-risk land-oriented occupations and activities
- Social interactions that facilitate frequent, informal, face-to-face contacts
- Close social or kinship relationships among community members
- Preference for informal support systems in times of need
- Individual and community self-reliance
- Small towns as centers of trade
- Churches and schools are centers of socialization for residents
- Importance of local health systems, especially hospital and longterm care facilities, to the local economy
- Overlapping roles and relationships for clinicians as community Members

Self-Reliant Behaviors

Self-reliance is another characteristic often attributed to rural residents. Historically, and today, the trait of self-reliance has helped families to survive in austere, isolated, and rugged environments. Self-reliance can help a patient through illness, with informal support from family, neighbors, or community groups; however, it can undermine healing if the patient and family avoid formal health care treatment and "tough it out." For example, a man abusing alcohol or chemical substances may avoid acknowledging his problem and from seeking much-needed professional services, due the enabling behaviors of his family. Or the community may view a woman with a mental illness as exhibiting a character flaw or weakness typical of her family, about which nothing can be done. Hence the admonition, "What happens in the family, stays in the family." To preserve family integrity, it becomes important to maintain secrecy, and not let others know about the problem (e.g., substance abuse, domestic violence, incest, rape, emotional disorders, stigma laden-illnesses, or unplanned pregnancy). Additionally, professional values and guidelines may conflict with family expectations, and coupled with community-defined standards of behavior or values, can impose stress for residents and the clinicians who care for them. Such scenarios can lead to patient-provider-family conflicts and, subsequently, ethics conflicts.

Rural Social Support Systems

The literature describes three levels of social support for patients in rural settings. The first level includes assistance volunteered by a patient's immediate and extended family, friends and neighbors. Although this type of support is not monetarily reimbursed, there is an expectation of reciprocity by those rendering the services. The second level includes support provided by a group or organization (e.g., civic organizations, homemakers' clubs, faith community, youth groups) in which members assist each other in the network during times of need (e.g., volunteering expertise, providing food, contributing financially, assisting with field work and farm chores).

Essentially these levels offer an "insurance policy" should a catastrophic event occur to anyone belonging to one of these community or faith groups. These care systems are a resource for rural residents who are coping with hardships and geographical isolation; yet they also reinforce perceptions of self-reliance. The third level of care consists of formal support, such as services provided by health departments, home health and hospice agencies, community nursing services, mental health centers, physicians, and hospitals. Financial remuneration is expected for these services, albeit in some cases on a sliding scale.

Urban residents often prefer the third level of support, since they may not have access to the informal care systems that characterize small towns. Rural communities, however, historically rely on the first two levels of support. In day-to-day activities, rural residents prefer to deal with someone they know ("kith and kin") rather than a stranger ("bureaucrat"), such as providing child care to a family during an illness, or preferring to receive care in the local hospital from the doctor and nurses who are neighbors, friends or relatives.

Rural Economics

Rural economics can impact a sick person's care-seeking behaviors. Generally, in small towns, salaries are low; there are limited benefits, including employee health insurance benefits, which results in a high rate of uninsured and under-insured. According to the Rural Assistance Center (RAC), "The uninsured in remote rural counties are not a peculiar sub-population of their communities: 68 percent come from families where there is at least one full-time worker; 30 percent are children; and almost two-thirds come from low-income families (less than 200 percent of the federal poverty level—less than $37,700 for a family of four). Families with two full-time workers, married couples, and the employed are also at greater risk of being uninsured if they live in a remote rural county.

The RAC also reports: "Remote rural residents are less likely to be offered health benefits through their employment: approximately 59% of workers in rural non-adjacent counties are offered employer-sponsored health insurance, compared to 69% of urban workers, and less than half of workers in rural nonadjacent counties are covered by their employers (compared to nearly 60% of urban workers)."

In addition to the economic issues that impact individual residents' use of health care services, rural health care facilities are a main economic driver in small communities. In most rural communities, the largest employers are the school system and the health care facilities. Because rural health businesses may experience frequent financial challenges, these financial issues are passed on and can significantly impact the broader community. For example, the closure of one rural hospital can create a dramatic ripple effect on the economic status of the community, many of whom may have had jobs there.

HEALTH STATUS OF RURAL RESIDENTS

Health and illness are defined in various ways, as influenced by individuals' cultural background. Among some rural residents, health is defined as "the ability to work; to do what needs to be done," reinforcing a work ethic. Thus, a patient with such views may not seek "formal" health care until he or she becomes too ill to work. Further, the practical limitations of time and work in a rural community mean that residents may wait to seek medical care until they can combine it with other business requiring a trip to town.

Health Disparities of Rural Areas as Compared with Urban Areas

- Higher infant and maternal morbidity rates
- Higher rates of chronic illnesses, such as hypertension and cardiovascular disease
- Higher rates of mental illness and stress-related diseases (especially among the rural poor)
- Lower rates of health insurance and pharmacy coverage plans
- Greater expenditures on prescription medications, associated with lack of pharmacy insurance benefits and/or more out-ofpocket costs for drugs because of lower insurance coverage
- Problems unique to rural occupations, such as machinery accidents, and skin cancer from sun exposure in farming or outdoor labor

A recent U.S. Department of Health and Human Services (HHS) report indicated that "rural residents are more likely to report fair to poor health status than urban residents, and are more likely to have experienced a limitation of activity caused by chronic conditions than urban residents." Rural residents have a higher prevalence of long term health problems as compared to urban residents; this is primarily attributable to rural areas having a higher proportion of poor and elderly residents, as well as more accidents and trauma.

Mental health needs appear to be particularly significant in rural settings where residents struggle with significant substance dependence, mental illnesses, and psychiatric-medical co-morbidity. It has been noted that suicide rates in rural areas have surpassed urban suicide rates for over 20 years. Yet, rural residents have relatively low mortality in light of their high rate of chronic illnesses. A serious gap exists in the health-status data of vulnerable and at-risk populations like those in rural areas. Knowledge of a community's demographics is particularly important, because health status can affect health care ethics conflicts. The HHS report concludes, "a scant provider network, lack of adequate and affordable health coverage, and difficulty accessing high-quality care can lead to worse health among rural populations."

Rural Health Care

Even in the most remote rural areas, communities have some health care system that includes formal organizations and providers, as well as informal support systems. Rural health care professionals work in a variety of settings, including those listed below.

- Federally-Qualified Rural Health Centers (FQHC, also known as Community Health Centers): FQHCs qualify for enhanced reimbursement from Medicare and Medicaid, as well as other benefits. FQHCs must serve an underserved area or population.
- Critical Access Hospitals (CAH): These are small (25 beds or less) federally designated facilities. CAHs provide essential services to a community and are reimbursed by Medicare on a "reasonable cost basis" for services provided to Medicare patients.*f*
- Rural Health Clinics: These must be rural and in a designated shortage area, eligible for cost-based reimbursement. The Rural Health Clinics (RHCs) program is intended to increase primary-care services for Medicaid and Medicare patients in rural communities. RHCs can be public, private, or non-profit. The main advantage of RHC status is enhanced reimbursement rates for providing Medicaid and Medicare services in rural areas. RHCs must be located in rural, underserved areas and must use midlevel practitioners, such as a nurse practitioner (NP) and physician's assistant (PA).
- Community Mental Health Centers: Federally enabled centers providing a safety net for mental health care in underserved areas.
- Private Practice Settings Clinics: Private clinics, providing primary care. Some private clinics are affiliated with CAHs.

Despite federal assistance, many of these various health care mechanisms continue to struggle economically. Low population density often means there is not a critical mass of consumers for a particular health care service; thus, rural facilities and services often are challenged financially due to the lack of a specialized-care revenue stream.

The contextual realities at these various health care settings have implications for health care professionals in rural areas. Clinicians in the rural context often are expected to function as generalists, because they care for individuals of all ages with a wide variety of health problems. For example, in one day at a rural hospital, a primary-care clinician may care for an obstetrical family (mother and newborn infant);

a patient with postoperative complications from recent surgery; and an elderly person, who has a life-threatening chronic health problem. Such diversity in a practice can be perceived as a positive benefit by many rural health care professionals, just as it can produce caregiver stress. To manage the inherent stress and ethics situations in rural settings, clinicians must be aware of both the formal and informal resources that make up the local health care system, as well as how the two interface, and what it takes to connect with these services.

Availability of Health Care Services

Availability refers to the existence of services and the presence of sufficient personnel to provide those services. Rural areas have fewer physicians and clinicians, nurse practitioners, and specialists; especially obstetricians, pediatricians, psychiatrists, and social-service professionals. Economically, a sparse population limits the number and array of services in a given region, as the cost of providing services to a few people may be prohibitive.

Access to Health Care Services

Accessibility refers to whether a person has the means to obtain and afford needed services, or is impaired by certain barriers to accessing health care. It has been noted that, "rural residents were more likely to report that cost had kept them from seeing a doctor than were urban residents. The proportion of adults who reported deferring care because of cost increased with the level of rurality." A person in need of health care can encounter various barriers ranging from economic to transportation barriers.

Barriers to Accessing Healthcare

- Long travel distances
- Lack of public transportation
- Limited telephone and Internet services
- Shortage of health care providers
- Limited economic resources—income and insurance
- Challenging roles and unpredictable weather conditions
- Inability to seek or obtain entitlements

Two other barriers which, like economic barriers, can limit access to health care include geographical and physical barriers. Consider the case of a rancher with a high income, who lives in a medically underserved frontier area, and who suddenly suffers a heart attack. In addition to economic and physical barriers to health care, there are cultural and educational barriers that result when rural individuals lack a particular skill or element of education. Perhaps a small clinic is seeking a grant to access funding to implement a health program. However, the staff is hampered because they lack grant-writing skills. Or there may be a community perspective that opposes the use of federal or state welfare programs. Rural-based clinicians must recognize the stigma attached to the use of some types of services, and then adapt delivery approaches to try and assure anonymity and confidentiality within a rural context where people tend to be acquainted or related.

Barriers to Accepting Healthcare Services

- Self-reliance leads to self-care for health issues (e.g., patients preferring to self-treat with over-the-counter medications, exercise, rest, or prayer)
- Beliefs about the cause of an illness and the appropriate healer for it (e.g., patients prefer to see a "medicine man/ woman," curandero, shaman, or clergyperson)
- Community values about illness (e.g., patients being stoic and suffering in silence rather than seeking care)
- Lack of knowledge about a physical or emotional disorder, or of the importance of formal services for prevention and treatment
- Difficulty in maintaining confidentiality and anonymity in a setting where most residents are acquainted

Acceptability of Health Care Services

Acceptability refers to whether a particular service is offered in a manner that agrees with the values of a target population. With the wide diversity among rural residents, acceptability of available services, like community nursing services. Geographic, demographic, social, and economic factors will impact a clinician's practice as well as the health status in a particular rural community. In turn, ethics conflicts in rural settings generally are associated with these characteristics.

Ethics Issues and the Rural Context

The features of rural lifestyle (geography, population density, cultural values, and health care systems and services) influence the ethics situations that rural clinicians encounter. Many of the previously described rural characteristics that influence ethics conflicts are listed below.

- Community values and beliefs that differ from professional standards
- Overlapping personal and professional roles within the community
- Threats to confidentiality and privacy
- Real and perceived geographic and professional isolation
- Limited access to, and availability of, health care services and providers
- Economic limitations, such as low income and lack of adequate insurance
- Clinician stress associated with community expectations, professional workload, and isolation
- Reliance on informal, non-professional community health care support networks
- Limited ethics resources

Overlapping Professional and Personal Roles

Probably the most common part of the rural context that can foster ethics conflict is the role of overlapping relationships between clinicians and patients. Due to the geographical and social structure of rural communities, rural health care providers commonly interact with members of the community in more than one relationship—i.e., a nurse may serve on a school committee which also has one of her patients on the board; a doctor's children may play with the children of his patient; a psychiatrist may attend a house of worship where some of her patients also go. Generally, rural providers live and work in the same place, and everyone knows one another. Everyone knows the community's physician and/or nurse, and it is difficult to escape that role. Rural health care professionals frequently interact outside the office with community members. These multiple relationships can enhance and complicate the patient-clinician relationship.

This regular contact allows rural clinicians to have a knowledge of their patients that is unlikely in more urban settings. Rural clinicians have the unique opportunity to understand their patients in depth, including the patient's personal values and perspectives. The patient-provider relationship is formed and cultivated in both the examining room and in the general store. This can be very beneficial for treatment; however, ethics issues can also arise in maintaining professional boundaries with patients. Almost every rural ethics issue encountered is influenced and shaped by provider familiarity and overlapping roles. Community Values Can Differ From Professional Practices

Authors have noted that rural residents from various cultures hold different views of pain, the etiologic explanations for sickness, tolerance of illness, and the use of folk healers. When the pervasive community values about illness differ from the traditional practice and ethos of clinicians, it is more likely that ethics conflicts will arise. Ethics conflicts may occur if providers show insufficient respect for cultural and community values. Recognizing community values and openly communicating with patients about how those values and beliefs differ from traditional health care clinical and ethics practices are key to the provider in addressing potential conflicts. Clinicians should also consider approaches to patient care that apply community values to their professional standards of practice. Cultural insensitivity can also exacerbate a mistrust of clinicians. More specifically, a clinician's attitude can impact the long-term health status of a patient who may be embarrassed about his or her health problem(s).

Embarrassment may be evidenced in certain ethnic or cultural groups, who may minimize symptoms of illness, not acknowledge self-care practices, or not seek care when it is needed. In such groups, health care is sought only for an acute illness or an emergency.

To diminish ethics issues related to cultural and community values, it would be beneficial for all health professional programs (medical training, nursing schools, technologist training, etc) to expose students to the rural environment and the cultural perspectives that they will encounter. For example, a rural-focused elective course could promote cultural competency in relationship to health care issues. Rural-based experiences and cultural understanding could go a long way to create a climate of mutual sensitivity and trust between clinicians and rural patients that could prevent health care ethics situations from occurring. Threats to Confidentiality and Privacy

Closely related to overlapping relationships are threats to confidentiality and privacy. Breaches in confidentiality and privacy can be both intentional and unintentional, due to the close-knit nature of rural settings. It is not unusual for rural residents to report that even though they are well acquainted with most residents in the community, they feel there is no one they can trust, or with whom they can discuss personal problems. This experience is in part attributable to small town residents' genuine interest in the lives of neighbors, friends and relatives.

Regardless of the setting, it can be devastating for those involved if their personal problems become public knowledge. Informal social structures, such as a church or extended family, in turn, can impose restrictions for those who desire to seek professional help for issues with moral overtones, such as drug and alcohol dependency, an unplanned pregnancy, sexuality issues, conflicts in personal relationships, or behaviors associated with mental illness. Further complicating the issues is the reluctance of some rural patients to access urban-based outreach services provided by professionals who are strangers, while services provided by the parish nurse may be openly welcomed.

Limited Access to Health Services

The limited availability, accessibility, and acceptability of health care services can foster ethics challenges that are regularly encountered in rural settings. The lack of services and specialists, such as mental health professionals, creates situations where the general practitioner may need to provide specialized mental health care. Other health care professionals, including nurses and social workers, may become the de facto primary-care provider in settings with limited resources. Many rural communities have no hospitals; yet travel to the large, distant medical centers can create burdens on both the patients and the rural providers. For example, a patient may need to travel long distances over challenging roads to find needed emergency care. Or a patient may need specialized cancer treatment at the distant medical center, but opts out of the care because of the distance and lack of support. Such situations place a great burden on rural primary-care clinicians who may be forced to provide care outside their areas of expertise.

Taken together, these factors (limited availability, accessibility, and acceptability) conspire so that small communities may have few health care services, lesser expertise, tremendous professional shortages, heightened ethical binds, and greater risk for errors and quality issues.

Community Economic Limitations

The economic status of rural residents can influence their use of health care services. A community member's willingness to accept needed care may be restricted by his or her economic situation. For example, patients requiring care may be underinsured or uninsured. Ethical questions can arise as a result of these situations. Rural health care professionals regularly encounter situations where they must decide whether to provide needed care with little or no reimbursement, which may potentially jeopardize both their patients' health and their own overall practice and standing within the community. Not providing needed services is a difficult decision, because it may be contrary to community values and the perceived role of the clinician. Because everyone in the community knows the physician, such decisions can never be made in a confidential manner. Resource allocation and access to health care are areas of great ethical challenge for rural physicians.

Isolation for a rural health care professional also assumes not having an immediately accessible network of peers who can provide support and consultation on a particular concern. Professional, and sometimes social, isolation may be associated with geographic isolation. Isolation can be perceived by the clinician as positive, negative, or a combination of the two. Some rural clinicians appreciate the opportunity for developing professional autonomy and creativity, while others find the expectations and responsibilities in rural practice to be stressful and overwhelming. Likewise, the lack of immediately available opportunities for establishing relationships with other professionals, or not having the central office nearby, can reinforce feelings of isolation.

Perceived professional isolation requires the clinician to evaluate and prioritize needs and types of services that can be provided to the local population. The clinician can experience considerable professional strain and stress from the lack of geographic access to other specialist providers, as well as the lack of continuing education, current telecommunications, and medical technology. Individuals who are uncomfortable when working alone, or who lack the confidence to make independent clinical decisions, probably would not fare well in a remote rural area. These individuals would likely be more comfortable in a less isolated setting with more support.

LIMITED ETHICS RESOURCES

Unlike their urban counterparts, many rural facilities and health care professionals have limited access to ethics resources to help them address ethics conflicts, including rurally focused ethics literature, ethics committees, and health care ethicists.

Limitations in rural ethics resources

- Rural-focused ethics literature
- Rural ethics committees
- Rural-based health care ethicists
- Rural-focused ethics training

To address these limitations, several strategies may help the rural clinician to identify rural ethics resources. Often the local health care facility is part of a larger network that has an ethics committee, an ethics consultant, and continuing education opportunities for participating members. Rural clinicians should ask and become informed about what is available to system members. Other small communities within a given geographical setting have combined resources and developed their own health care ethics committees and educational resources. In some cases, health care providers in small communities have partnered with institutions of higher learning to educate local health professionals about ethical principles and decision-making. It is important that rural health care providers consider, though, that educational programs for health professionals generally are located in urban areas, and most clinical experiences occur there. In turn, students are not necessarily exposed to rural patients and rural health care systems. In these cases, rural clinicians must become proactive and educate urban-based educators about rural contextual features that can impact ethical situations occurring in more austere and remote environments. Potential ethics resources

- Local health care facility's ethics committee
- State network of ethics committees
- Multi-facility ethics committee
- Academic medical center ethics program
- University or college ethics department
- On-line resources offered by health professional organizations.

★★★

12

Rural Urban Migration

INTRODUCTION

Migration is considered as the movement of people from one geographical region to another, which may be on temporary or permanent basis. People migrate based on the prevailing conditions and the reasons for it vary from one person to another depending on the situation that brought about the decision. Migration is a selective process affecting individuals or families with certain economic, social, educational and demographic characteristics.

Migration is a complex phenomenon and can be better understood only by a probe into the migration process and the consequences of migration. Migration process covers decision on migration and job search. Of the various types of migration streams rural to urban migration represents mainly employment migration. The increasing degree of industrialization , consequent increase in demand for labour, existence of informal sector activities, scope for self-employment and above all the preparedness to accept any kind of job for an earning drive people towards urban areas.

Migration occurs as a response to economic development as well as social, cultural, environmental and political factors and effects on areas of origin as well as destination. People tend to move away from a place due to need to escape violence, political instability, drought, congestion in various dimensions and suspected or real persecution. Also, adverse physical conditions such as flood, landslide (erosion and earthquake), insects and pests, soil infertility contribute largely to the reasons why people leave one environment for another.

Migration is a two stage process. The first stage relates to the decision to move and the second stage to the choice of destination. Once the destination is chosen, the decision on the manner of search for the first employment is important. From the destination the types of linkages the migrants want to maintain with their places of origin through visits and remittances is also equally important for the study of the consequences of migration. The in-depth micro level studies are important for getting the guidelines relating to the nature and relevance of macro level measures for application at the appropriate time. The aggregate behaviour after all is governed by the characteristics, attitude and behaviour of individual migrants.

Studies by Okpara (1983) reveal that migrants from rural to urban areas are always many compared to those coming from urban to rural areas. The movement from rural to urban areas makes a negative impact on the quality of rural life especially when such migrants carry away their needed consumption into the city. Migration of young adults from the rural areas also placed a greater burden on the farmers. For farmers to cover the same area of land as when he had extra assistance, he must work much longer hours thus depriving him of some time for leisure or participation in various social activities. On the other hand, studies by Ijere (1994) reveal that rural–urban migration has a positive impact on urban growth and social development, which makes generation of employment opportunities and provision of educational facilities and transportation infra-structure for the migrants.

Today, urban-rural migration is one of the most important modes of migration which exist in the society. Until recently, researchers have not paid much attention to the rural –urban drift. Studies by William (1970) reveal that some factors could send one off a place, which might be due to crisis, old age, transfer, retirement and invasion of pests and diseases. Also some social factors may evolve to make people migrate to rural areas. Studies by Jibowo (1992) show that factors influencing people to migrate to rural areas might be as a result of city congestion, traffic, dams, sanitation problems, increasing urban unemployment, increased crime rate and housing problem. These problems are so great in many developing countries that rural develop-ment has been a cardinal programme of political campaigns by many politicians.

This study therefore investigated those factors that are associated with the urban to rural migration in Nigeria. Specifically, this study investigated factors associated with movement of people from towns to rural areas and some of the socio-economic characteristics of the migrants. The study hypothesized that there are no significant relationships between lengths of stay of migrants in the rural areas and some of their socio-economic characteristics such as age, gender, marital status, years of formal education and occupation.

Socio-Economic Factors Influencing Migration: The results of the study show that a majority (67.5%) of the respondents left the town for rural areas due to factors ranging from inability to secure jobs, transfer from their place of work, retirement to high cost of living in the town. The greatest proportion of the migrants moved to rural areas due to high cost of living in urban centres. Another 11.3 percent moved away because of congestion in the urban centres.

Majority of the respondents are married men who had received some forms of formal education. A majority of the migrants moved to rural areas in the recent past. The major factors that influenced the movement to rural areas from the urban centre included inability to secure jobs and high cost of living in the urban centres, transfer to rural areas in their places of work and retirement. The ages and marital statuses of the migrants contributed significantly to their lengths of stay in the rural areas.

It is recommended that government should encourage the drift of people to rural areas by making available such amenities that would encourage the stay of the migrants. Also, migrants should be encouraged to form themselves into cooperative societies which can be used to attract the attention of the Government or Non-governmental organizations to establish different cottage industries in the rural areas. This encouragement would enhance rapid rural development and bring the rural communities to the mainstream of national development.

The study of migration in general and rural-urban migration in particular has for long been an important area of research in development economics. A large body of literature has grown up in recent years around the topic in contemporary less developed In a country with 600 million farmers, of which 40 percent are willing to quit farming for various reasons, mass migration from rural to urban areas has increased rapidly. Between 1991 and 2001, 73 million people have migrated from the rural areas to elsewhere. Mass migration is a phenomenon that is a consequence of various problems in the rural India.

Prime Minister, Manmohan Singh, suggested that if people from the rural areas are brought to urban areas, the state could serve them better. But, this is a wrong notion which can lead to the destruction of agriculture on the whole, almost or completely damaging *"Bharat"*.

There are umpteen number of reasons for this, namely: health, poverty and hunger, water and sanitation, women and children, education and employment, environmental problems, resources, industries and corporates, etc. Now, let us analyze each reason and how and why the proposal cannot turn into action.

In terms of health, India has one of the most neglected health care systems in the world. India's ranks among top five countries in the world with most number of HIV positive cases, present IMR in India is 52 deaths per 1000 live births and MMR is 230 per 100, 000 live births according to WHO statistics. Though, private and corporate hospitals are blossoming in the cities, villages in rural India still suffer from lack of proper health care services. Setting up public health centres (PHC) and community health centres (CHC) have solved a few problems in the villages. However, lack of proper equipment, operation theatres, paramedics, technicians and doctors are leaving most of these health centres in shams.

Services like 108 have been introduced in villages in case of emergency, but the patient will be driven in a van through roads which almost kill the person before reaching the PHC or the CHC which is mostly located in towns, about an hour drive from any village. Population is another major problem in our country. In the name of family planning, forced sterilization, birth control pills and Intra-uterine devices (IUD) are used. The effects of these can cause hormonal problems in women or sometimes, death.

Once we probe deeper into issues related to health, hunger and poverty stand tall in the list of factors affecting the health of people. India ranks 63rd in poverty Index and its rank in the standard of living compared to that of 186 countries is 126, which is much lower than many other developing countries. More than 320 million people go hungry to bed every day. Of these, many die, many suffer from various diseases. Most of them among the 320 million hungry people are the ones who live on less than Rs. 20 a day.

Apart from food security, what we also need to look at is water and sanitation problems. Even today, many people in the rural areas walk 10 kilometers or more to get water from the nearest drinking water sources. They are infected with various diseases because of sanitation problems. Depletion in ground water level and water pollution due to prevalence of toxics, sewage and other pollutants is also a trouble. For example: Many districts in Karnataka like Kolar have high fluoride content in them which makes the bones of people brittle, causing joint pains. Apart from this, fluorosis also decreases fertility rate in men and women. The reason for high fluoride content is because of depletion in the ground water level.

ROLE OF MIGRANT

A migrant as a job seeker and a decision maker has to acquire and use information to take rational decision in an ever changing and uncertain environment. Job search becomes an integral part of decision to migrate. Job search becomes a continuous process in the case of those who want to keep on changing with the interior motive of moving up in the occupational ladder or moving up in the economic status. In recent times, the focus of attention of migration study shows a shift from individual to family and family welfare maximization. Family as a unit has been recognized and due importance has been assigned to family considerations in influencing migration decisions. The economic studies of Polacheck and Horvath (1977), Mincer (1978), Stark (1978 & 1979) and Banerjee (1986) have discussed migration in the context of family.

Migrants from rural areas residing in the city exhibit a particular behavioural pattern through the linkages with their origin. In some cases, there is the deliberate link maintained. In all kinds of linkages the self-interest combined with maximization of family welfare is the dominant force. Despite geographical separation, sentiments and affectionate ties between family members are not weakened and the close bonds between the different components of the family are reflected in regular transfer of resources, frequent visits and occasional movement of persons from one unit to the other. A wife and family in the origin is obviously a strong incentive to visit as well as to remit money income. In a few cases the strong desire to go back to the area of origin is also a noticeable factor.

Permanent return may be caused by retirement, fulfillment of a target, growing rural opportunities etc. Dissatisfaction in the earnings and the inability to support the family in the rural area etc, may lead to permanent migration.

The behaviour of migrants before and after migration could be well understood only with a proper analysis of the factors governing the same. In predicting how different migrants with different responses to the existing conditions will behave, it is necessary to know something about their specific characteristics like age, education, martial status, costs to be incurred and economic characteristics like employment and property status. Pre- migration search behaviour is very important for it might provide scope for shaping the future of the potential migrants. Personal characteristics and a number of related factors govern the pre-migration behaviour. An analysis of factor governing this type of search might be more useful and interesting.

The urban search represents post- migration behaviour and reveals nature, attitude and aspirations of those who prefer urban based search initially. Focusing attention on the determinants of rural based search would naturally throw light on the determinants on urban based search. This in turn is expected to throw light on some of the essential aspects of the rural migrant behaviour in the urban area. The important characteristics that can be associated with rural based search and sources of information appear to be having a dominant influence on the decision regarding the type of search.

The importance of the accompanying characteristics cannot be minimized. Those who belonged to the older age group of 30 and above at the time of the migration and married migrants had shown preference for rural based search. Those who had migrated earlier, educated, propertied and professionals, administrators and executives had preferred search in the urban area.

Those who belonged to the clerical category had to satisfy themselves with the search in the area itself. The overall picture shows that those who belonged to the lower category could not afford to have search for job in the urban area. However, the marginal difference revealed from the table based on the research study area does not help drawing categorically any conclusion regarding preference for one type of search over the other. But however, there is a clear indication that personal characteristics, length of stay and property position influence the basic attitude of the potential migrants and their behaviour.

In the urban area, search for the first job as well as the subsequent jobs represent the post-migration behaviour. Urban search seems to be a continuous process in the case of those who want to move up the occupational ladder or those who want to shift to other jobs for improving their economic status. Job shifts might be within the same occupation and industry or might be totally a different job and industry. Job mobility represents movement from one employer to another and also encompasses movement among jobs within the same firm. The non-wage earners include own account workers namely self-employment and unpaid family workers.

The post-migration behaviour in the urban area representing linkages, the migrants maintain with their areas of origin, take the form of visits and remittances and to some extent the probable return migration also. Visits in some cases appear to be regular while in others it is irregular. Even migrants who have no family and property at the origin tend to visit their places. As for remittances, naturally family welfare considerations are important. The desire to go back to the place of origin necessitates not only the remittances but also a change in the size of remittances depending upon the link between the migrants and the others at the origin.

As for the return migration, the study brings out the negative relationship with age indicating thereby those who belonged to the old age have no inclination to return to the area of origin and those in the younger age group have plans to get back to their place of origin. Those engaged in permanent jobs indicate least inclination to go away either in the present or in the future. Education level does not seem to have any special impact on the desire. The influence seems to be similar at all levels. Probably, the desire is for a few migrants to go back to their origin and settle down peacefully. The initial tendency to stay for a longer period is mainly to get the best out of the city life.

As mentioned earlier, the important aspects of migrant behaviour that need consideration for any discussion on the consequences of migration in developing countries are the pre-migration and the post migration behaviour of the individual migrants, that is, the nature of their search behaviour and contact they want to have with their areas of origin after going into the urban area. The former could be taken to represent the cause induced behaviour of migrants and the latter, effect induced behaviour of migrants. With the strong desire for rural based search and the necessary patience for getting specific information and the possibility of a certain percentage of migrants joining the informal sectors, the problem in the urban area could be minimized to some extent. The increase in the number of dependents in the urban area definitely adds to the problems not only to the individuals but also of the entire urban area. Generally whole time search undertaken by the migrants might bring fruitful results in a short time. Part time search would be successful in the short period only when there is access to information about specific jobs, additional skill or training gained during the first job period. The search behaviour seems to emphasize that there is a definite relationship between various socio-economic variables and other external factors on the migrant behaviour.

Highly qualified workers might show a tendency for job shifts. The gain in skill and experience might induce some workers to take steps to move up in the occupational ladder. Rural migrants being mostly semi skilled and less educated might not be in a position to take up jobs requiring skill and being generally poor might not be in a position to start business involving more risk and large funds. The self-employment category might be contributing to the area positively only if larger investments are accompanied by larger earnings.

The links maintained by the migrants with their areas of their origin tend to benefit rural area. The visits either frequent or even occasional, develop an interest in the minds of the migrants in the development of the areas concerned. Remittances indicating cash flows into the rural areas with specific purposes help the areas and the people in all possible ways. Remittances might offer the necessary stimulant to the growth of agricultural production and technological change. Remittances are significant inducing socio-economic development of the areas of the origin and also the distribution of income both within and between areas of origin and destination.

The earning prospects that urban areas promise for semi-skilled and untrained labour from the rural areas will continue to strengthen the migration tendency without of course having an unfavorable impact on labour market situation. The urban areas offer a wide scope for self-employment activities of varied nature and this enables the migrants to be content with whatever activity they first chose to do. So long as strong family binding is present, the

permanent settlement in the urban area need not be dreaded. Though remittances are not always used for productive purposes in the rural areas yet they help in the dissemination of information and encourage linked migration. The fact that migration leads to overcrowding and pose problems of basic amenities in urban areas cannot be totally ignored. But the fact remains that so long as the conditions are favourable for the migrants in the urban areas, the rural areas from where individuals migrated get the advantages. This in turn could reduce the intensity of the problems on the urban areas.

RURAL-URBAN MIGRATION AND ECONOMIC GROWTH IN DEVELOPING COUNTRIES

Rural-urban migration has long been associated with economic development and growth in the economic literature. In particular, Todaro and Harris-Todaro-type probabilistic models that examine migration have concentrated on the expected wage disparities between rural and urban (formal) labor markets as a driving force behind migration decision. These models, which are static and partial equilibrium in nature, have virtually ignored the cost-of-living differentials across regions that arise from the presence of regional non-traded (home-) goods. Moreover, even in dynamic general equilibrium models, equations specifying labor market clearing conditions have neglected to recognize a missing endogenous variable, the households' choice of residency, and the corresponding equations necessary to cause the labor market to clear as well. Effectively, adding these conditions to the model allows agents to move from one region to another and to bring their utility function and budget constraint with them to the new region of residency. This condition profoundly affects the spatial distribution of economic activity. Furthermore, when factor market imperfections are modeled, e.g., the segmentation in labor and capital markets across regions, these factors earn different rates of return thus greatly influencing the pattern of spatial economic development. The main objectives of this paper are to model the residency choice decision in the context of a dynamic general equilibrium economy, to identify the channels through which segmentation in capital markets in developing countries induces migration from rural to urban regions, and to explain how uneven economic growth may emerge as a consequence. This paper incorporates cost-of-living and income differentials across regions into the migration decision of households in a dynamic general equilibrium setting. With the use of a dynamic general equilibrium model, we can capture the migration pattern as a response to changes in cost-of-living, as well as to the evolution of real wage differentials as capital accumulates due to household savings and as the rural-urban production sectors respond to the Rybczynski-like effects of competition in factors of production.

Using a model that extends the standard Ramsey-type growth model, we investigate the endogenous pattern of migration in a developing country economy in the process of economic growth and structural change. The standard Ramsey-type growth model is thus extended to include two types of households in a regional, multi-sectoral environment with capital market segmentation. In particular, to best assess the impact of capital market segmentation on the economy as a whole and on specific macroeconomic variables, a policy experiment is conducted under the cases of with and without capital market segmentation: when a policy's shock is introduced, the economy's performance, as well as migration patterns are examined when there is segmentation in capital markets, and when there is a perfect capital market. The policy experiment is conducted by lowering the labor tax rates levied on the employers in the urban formal sector. The model is calibrated to Turkish economy for the year 1997, which has a large rural population at about 42 percent of the total population as of that year. Data are compiled from Turkish National Accounts Statistics and Labor Statistics.

Initial results from numerical simulations show that in a model economy with a large rural population and segmentation in its capital markets, a policy change in the economy such as reducing the labor taxes imposed in the urban formal sector induces migration from rural to urban areas, and this migration continues along the transition path to a new long run equilibrium. Large drops in output in rural areas are detected, whereas the output in the urban region grows along the transition path. However, the same economy reacts to the same policy change much differently after it undergoes an institutional reform such as the integration of its capital markets. As the economy adjusts to a new equilibrium once a policy change is introduced, relative to the case with segmented capital markets, no large changes in the macroeconomic variables occur. Especially, rural households choose to remain in the rural region.

RURAL-URBAN LINKAGES

What is 'Urban' and what is 'Rural'?

The difference between urban centres and rural areas may seem so obvious that definitions should not be an issue. However, there can be major variations in the ways in which different nations define what is an urban centre. The criteria used include population size and density, and availability of services such as secondary schools, hospitals and banks. However, the combination of criteria applied can vary greatly. Even the population thresholds used can be different: for many African nations it is 5,000 inhabitants, while for most Latin American and European nations it can be as low as 2,000 or 2,500 or even just a few hundreds inhabitants.

This wide fluctuation in definitions has three important implications.

- Official classifications should be treated with caution – for example, a large proportion of settlements classed as 'rural' in China and India would fall within the 'urban' category if they used the criteria and population thresholds adopted by many other countries. Given the size of the population of these two countries, this would significantly increase the overall proportion of urban residents in Asia and in the world.
- International comparisons are difficult, as they may look at settlements which, despite being classed in the same category, may be very different in both population size and infrastructure. In addition, the reliability of data on urbanization trends within one nation can be compromised by changes in the definition of urban centers over time.
- Public investment in services and infrastructure tends to concentrate on centers that are defined as urban. As a

consequence, investment can bypass settlements not defined as urban even if these can, and often do, have an important 'urban' role in the development of the surrounding rural areas. Within national and regional urban systems, larger cities also tend to be favored with public investment over small and intermediate-sized urban centers, including those with important roles in supporting agricultural production, processing and marketing.

Outside the City Boundaries: The Peri-urban Interface

The physical boundaries of urban built-up areas often do not coincide with their administrative boundaries. The areas surrounding urban centers generally have an important role in providing food for urban consumers, with proximity lowering the costs of transport and storage. It is difficult to make generalizations on the nature of peri-urban areas, which depends on the combination of a number of factors including the economic and infrastructural base of the urban center, the region and the nation; the historical, social and cultural characteristics of the area, and its ecological and geographical features. Peri-urban areas around one center are also not necessarily homogenous: high- and middle-income residential developments may dominate one section, while others may host industrial estates and others provide cheap accommodation to low-income migrants in informal settlements.

The peri-urban interface around larger or more prosperous urban centres is also the location where processes of urbanisation are at their most intense and where some of the most obvious environmental impacts of urbanisation are located. They are often characterised by:

Changes in land use: land markets are subject to competitive pressure as urban centres expand and speculation is frequent. Whether low-income groups such as small and marginal farmers or residents of informal settlements can benefit from these changes, or end up losing access to land, depends largely on land rights systems.

Changing farming systems and patterns of labour force participation: because peri-urban agriculture can be highly profitable, small farmers may be squeezed out by larger farmers who can invest in agricultural intensification. As a consequence, wage agricultural labour often becomes more important than small-scale farming, attracting migrant workers. On the other hand, residents of peri-urban areas may benefit from employment opportunities in the city.

Changing demands for infrastructure and pressure on natural resource systems, with many rural dwellers' access to resources having to compete with urban demand (for example, for water, fuelwood and land for non-agricultural uses) or affected by urban-generated wastes.

Variations in the characteristics of peri-urban areas can be important. For example, in the growing number of extended metropolitan regions in Southeast Asia, agriculture, small-scale industry, industrial estates and suburban residential developments co-exist side by side. Availability and affordability of transport are essential for the intense movement of goods and the extreme mobility of the population. In other contexts, and especially in less industry-based economies such as many countries in sub-Saharan Africa, agriculture still prevails in peri-urban areas although often with significant shifts in land ownership and use. This is especially the case where smallholder productivity is low because of the increasing costs of inputs and limited credit availability. Other problems include poor access to urban markets due to a lack of roads and physical infrastructure and the tight control over access to the urban market-places by middlemen and large traders. Thus, despite proximity to urban consumers, small farmers may be easily squeezed out, especially as the value of land in peri-urban areas increases with the expansion of the built-up center.

The Extra-urban Impact of Urban Activities: Cities' Ecological Footprints

Most cities draw heavily on their surrounding regions for freshwater resources. Most urban wastes end up in the region surrounding the city, for example, solid wastes disposed of on peri-urban land sites (either official or illegal) and liquid wastes either piped or finding their way through run-offs into rivers, lakes or other water bodies close by. Peri-urban areas may also be affected by urban air pollution.

Understanding Rural-urban Differences and Rural-urban Linkages

There is a need for an understanding of development that :

- Encompasses both rural and urban populations and the inter-connections between them;

- Acknowledges that where people live and work and other aspects of their local context influences the scale and nature of deprivation (whether they live or work in rural or urban areas); and
- Recognizes that there are typical 'urban' and 'rural' characteristics that cause or influence people's livelihoods, although care is needed in making generalizations because of great diversity between different urban locations (and rural locations).

These should be regarded as two ends of a continuum with most urban and rural areas falling somewhere between these extremes. The text noted earlier the importance of non-farm income sources for many rural households (including remittances from family members working in urban areas) and the importance of agriculture and/or of rural links for many urban households (including urban centers with many residents who work seasonally in rural areas).For all the contrasts between 'rural' and 'urban' highlighted in the figure, there are many exceptions. It is also useful to see in the middle of the continuum between 'rural' characteristics and 'urban' characteristics a 'rural-urban' interface in which there are complex mixes of 'rural' and 'urban' characteristics. For instance, many of the areas around prosperous cities or on corridors linking cities have a multiplicity of non-farm enterprises and a considerable proportion of the economically active population that commute daily to the city or find work seasonally or temporarily in urban areas. Many rural areas also have tourist industries that have fundamentally changed employment structures and environmental pressures.

Governing Across and Beyond the Rural-urban Boundary

If well managed, the interactions between towns and countryside are the basis for a balanced regional development which is economically, socially and environmentally sustainable. Local development is increasingly associated with decentralisation processes, on the assumption that local government is 'closer' to citizens – meaning that it is both more accountable to them and that it has a better understanding of local needs and priorities. With regard to rural-urban linkages, locai government can play an important role in facilitating positive interactions and limiting negative exchanges:

- It is best placed for decision-making on physical transport and communication infrastructure; however, expenditure for infrastructure can be significant and well beyond the means of local government. Wider alliances, which increase access to financial resources, are therefore necessary.
- The management of natural resources and wastes is an important area of local government intervention. However, it often includes much wider areas than those administered by local authorities, and requires alliances with other local, regional, national and sometimes cross-border governments.
- National level policies also have an important role, for example, with respect to access to land and land ownership and titling in both rural and urban areas. Clearly, this is not the responsibility of local authorities but is nevertheless crucial for local development planning and practice.In short, understanding rural-urban linkages matters because it provides the basis for measures that can improve both urban and rural livelihoods and environments. Ignoring them means that important opportunities will be lost, and in many cases it will also contribute to poor and marginal people's hardship. There are urban initiatives that can reduce ecological damage to rural areas, and help support regional development. However, with a narrow urban-centric approach, such initiatives are unlikely to be given the priority they deserve.

IMPACT OF RURAL-URBAN MIGRATION ON THE SUSTAINABILITY OF CITIES

The process of urbanization is now more rapid and massive and affects a greater part of the world than ever before, mainly because it is now rampant in the less developed countries, which still board 3/4th of the world's people. The migration of hundreds of millions of rural folk to cities in these still chiefly agrarian countries is revolutionizing the life of humanity just as surely as are the other major aspects of economic and social modernization. The unprecedented rates of over-all population growth are helping, along with the rural-urban migration, to swell the populations of individual cities more than ever before. Necessarily, social, economic, and political problems of major significance are being created by the huge rural-urban migration and the rapid rise of megapolises in countries whose main orientation has until recently been agricultural.

This topic attempts at understanding the challenges posed by Indian cities on their path to sustainability. This paper uses primary as well as secondary research for working. The paper shows: (a) how the rural regions affect the sustainable development of cities through migration and in retrospect also ruin their own virility and prospect of sustainable endeavour and (b) Some models to use in generating a solution to the issue – to create a balance in urban and rural development for sustainability at both the ends.

Preliminary Overview

For many years, rural's urban migration was viewed favourably in the economic development literature. Internal migration was thought to be a natural process in which surplus labour was gradually withdrawn from the rural sector to provide needed manpower for urban industrial growth process. This was deemed socially beneficial because human resources were being shifted from locations where their social marginal product was often assumed to be zero to places where this marginal product was not only positive but also rapidly growing as a result of capital accumulation and technological progress. This process was formalized in the Lewis theory of development. However, as Richard Jolly states:

Far from being concerned with measures to stem the flow, the major interest of these economists (i.e., those who stressed the importance of labour transfer) was with policies that would release labour to increase the flow. Indeed, one of the reasons given for trying to increase productivity in the agricultural sector was to release sufficient labour for urban industrialization. How irrelevant most of this concern looks today (1970).

In contrast to the pro-migration viewpoint, Indian research and experience has made clear that rates of rural's urban migration have greatly exceeded rates of urban job creation and swamped the absorptive capacity of both formal-sector industry and urban social services. Migration can no longer be casually viewed by economists as a beneficent process necessary to solve problems of growing urban labour demand. On the contrary, migration today remains a major factor contributing to the phenomenon of urban surplus labour; a force that continues to exacerbate already serious urban unemployment problems caused by the growing economic and structural imbalances between Indian urban and rural areas. Migration exacerbates these rural's urban structural imbalances in two direct ways (Todaro M. 1997): First, on the supply side, internal migration disproportionately increases the growth rate of urban job seekers relative to urban population growth, which itself stands at historically unprecedented levels, because of the high proportion of well-educated young people in the migrant system. Their presence tends to swell the urban labour supply while depleting the rural countryside of valuable human capital. Second, on the demand side, urban job creation is generally more difficult to accomplish than rural job creation because of the need for substantial complementary resource inputs for most jobs in the industrial sector. Moreover, the pressures of rising urban wages and compulsory employee fringe benefits in combination with the unavailability of appropriate, more labour-intensive production technologies means that a rising share of modern-sector output growth is accounted for by increases in labour productivity. Together this rapid supply increase and lagging demand (what many now refer to as jobless growth) tend to convert a short-run problem of resource imbalances into a long-run situation of chronic and rising urban surplus labour (Todaro M. 1997).

Rural India

Until a recent past, with 80% of India in rural regions, it was reasonable to say that Villagers in India manifest a deep loyalty to their village, identifying themselves to strangers as residents of a particular village, connecting back to family residence in the village that typically extends into the distant past. A family rooted in a particular village does not easily move to another and even people who have lived in a city for a generation or two refer to their ancestral village as our village." But today, the scenario has changed" migration has obliterated this factor of life from villages and, this new trend of greater urbanization has created a profound social, environmental, political and economic dilemma for all segments of society. Rural poverty is bad enough, but its problems are compounded when families leave their rural homes to seek a livelihood in overcrowded city slums, leaving behind deep-rooted traditions and ties to the extended family and the village seniors.

For India, development programs have been geared towards economic growth (as measured by GNP or per capita income). Once economic growth occurs, the planners believe, it is followed subsequently by many other changes in the areas of demography, stratification, polity, education and family (Chandrasekhar 1972; Dandekar and Rath 1971). It is with these hopes that immediately after independence, India, a young country and a large and promising democracy, moved in the direction of modernization. The first five year fiscal plans were totally geared to industrializing and developing the economy (Desai 1972; Davis 1968). Consequently, investment in economic growth has been biased toward the capital intensive urban centers, despite the fact that 80 % of India resides in the rural areas. The poor from the rural areas have no other option but to seek a livelihood in the urban centres. The rural-urban imbalance in development provides an explanation for the unprecedented growth of urban centers and slums (Srinivasan, 1997).

The Indian Planning Commission has targeted 8% growth for the Tenth Five Year Plan. Agriculture, the mainstay of villages, cannot grow faster than 2% at the most (Mukherji S., 2002). So, it is inevitable that either, (a) villages get relatively impoverished or (b) massive rural-urban migration results. The probability is that both will happen. At this rate, if no corrective steps are taken, in another 20 years, rural incomes will be 1/8th of urban ones or urban populations will quadruple, with slum populations going up by an order of magnitude (Mukherji S., 2002). Either way, the current growth paradigm is not sustainable either socially or politically.

In the developed countries of the West, the rural to urban migration and the urbanization are associated with a vertical shift in the labour force from the agricultural sector to the urbanised-industrial sector, whereas in the developing regions as India, migration is from rural agricultural sectors to urban informal sectors. Mukherji says that this is a typical manifestation of under-development, poverty, and spatial disorganisation of economy of underdeveloped sectors of the society, which arose partially as a result of past colonisation and its adverse consequences on space economy (1981). These displacements are not due to structural changes within the labour force as seen in the West, but as a dislocation of uprooted workers and peasants from the marginalised countryside to involuted urban centres. This displacement in India is a typical characteristic of urban growth that has outpaced industrialisation. It is a system of underdevelopment, and it tends to compound further underdevelopment (Mukherji S. 2002)

Urban India

The Indian urban population rose from a small figure of 25.6 million in 1901 to 212.8 million in 1991 (Census, 2001) that is 26% of the total population – urbanised. Many problems could arise due to such heavy influx of population and the immediate ones that show up on the radar are socio-economic and environmental problems. The UN says that if urbanization continues at the present rate, then 46% of the total population i.e. 634 million people will be in urban regions of India by 2030 (UN, 1998). If such an exodus is not checked and corrected, it would lead to extreme urban decay and malice where urban sustainability would then become one big utopian vision.

Pattern of Movement: Large groups of landless, unskilled, uneducated, illiterate labourers and petty farmers leave their villages and go to distant large towns or cities like Bombay, Delhi and Madras. They do not go to their neighbouring smaller town centres or districts because these regions have already reached their saturation point and fail to provide even the minimum services to the migrants. The only alternative that is left to the migrants are the large unknown metropolises only seen in movies as glamorous beds of comfort and wealth. Although the realisation is bitter, their arrival causes undue stress to the urban system which is now to deal with pavement dwelling, slums, squalors, disease, crime, and ultimate degeneration. The urban system breaks down often due to these stresses water shortage, electricity, transportation, education, housing, security and other services fail. When migration into urban regions continues, ongoing and increasing demands are made on the local economy which is anyways struggling for sustainability. This causes a major drain on economy. In some cases, drains on public funds do not stop at local coffers but instead invade national revenue pools, thus forcing adjustments in national priorities. Of course, this last scenario, where it occurs, is not the only way in which national development potential can be impacted by the urban influx. Since metropolitan areas house the strengths of national economies, the efficiency of those economies is impeded by anything that impairs the efficiency of the city (David McKee, 1994).

An ardent observer of the urban scene, Frank Lloyd Wright was quite skeptical of the viability of the twentieth-century metropolis. In a medical analogy, he suggested that, "To look at the cross section of any plan of a big city is to look at something like the section of a fibrous tumour" (Wright, 1958). Broadening his analogy, he maintained, "In the light of the space needs of the twentieth century we see there are not only similar inflamed exaggerations of tissue but more and more painfully forced circulation: comparable to high blood pressure in the human system". And as colourful as Wright's analogies may seem, there is little doubt that many major metropolitan areas face problems of congestion and over-concentration. That being the case, various planning procedures and urban renewal schemes may be out of step with actual needs.

In recent years, most advanced economies have become service oriented. That being the case, it should not be surprising that large cities have become service oriented as well. Traditionally, it may have been the industrial or manufacturing sector which played a larger causative role in city size than did service activities. However, the latter appear to have become more influential in determining the physical growth patterns of large metropolitan complexes in advanced economies (David McKee, 1994). Despite such problems, it appears as though the processes of economic development and urbanization are inextricably linked. Unfortunately, the actual processes of growth and change in Third World metropolitan areas have received surprisingly little attention. "Cities grow in size as a natural consequence of improvements in agriculture and the expansion of industry" (McKee and Leahy, 1974). It has been suggested that dual labour markets exist in the urban areas of the Third World "not just between agriculture and manufacturing" (Tisdell, 1990). This duality in urban labour markets may have the potential for impeding development. Whether such an impediment is strictly local in nature or wider in scope depends upon how it impacts the needs of the modern sector and how government priorities must be adjusted to deal with the negative externalities of dualism in an urban setting. Neoclassical thinking implies that as surplus labour enters the urban employment market, it must be absorbed if ongoing economic progress is to be achieved. Unfortunately, many fledgling urbanites lack the skills needed to function in their new environment; thus, their arrival "merely relocates the surplus labour problem in a setting which makes it more difficult to control" (McKee and Leahy, 1974). When natural increases in urban populations are considered along with in-migration, the labour absorption issue takes on even more serious proportions.

Informal Sector

A major focus of development theory has been on the dualistic nature of developing countries' national economies the existence of a modern, urban, capitalist sector geared toward capital-intensive, large-scale production and a traditional, rural, subsistence sector geared toward labour-intensive, small-scale production. In recent years, this dualistic analysis has also been applied specifically to the urban economy, which has been decomposed into a formal and an informal sector (Todara M. 1997). The bulk of new entrants into the urban labour force seem to create their own employment or to work for small-scale, family-owned enterprises. The self-employed were engaged in a remarkable array of activities, ranging from hawking, street vending, letter writing, knife sharpening, and junk collecting to selling fireworks, engaging in prostitution, drug peddling, and snake charming. Others find jobs as drivers, cabbies, mechanics, carpenters, small-scale artisans, barbers, apprentices, and personal servants (Todaro M. 1997). Given the unprecedented rate of growth of the urban population in developing countries that is expected to continue and the increasing failure of the rural and urban formal sectors to absorb additions to the labour force, researchers are devoting more attention to the role of the informal sector in serving as a panacea for the growing unemployment problem. The informal sector is characterized by a large number of small-scale production and service activities that are individually or family owned and use labour-intensive and simple technology. Such enterprises tend to be operated like monopolistically competitive firms with ease of entry, excess capacity, and competition driving profits (incomes) down to the average supply price of labour of potential new entrants. The usually self-employed workers in this sector have little formal education, are generally unskilled, and lack access to financial capital. As a result, worker productivity and income tend to be lower in the informal sector than in the formal sector. Moreover, workers in the informal sector do not enjoy the measure of protection afforded by the formal modern sector in terms of job security, decent working conditions, and old-age pensions. Most workers entering this sector are recent migrants from rural areas unable to find employment in the formal sector. Their motivation is usually to obtain sufficient income to survive, relying on their own indigenous resources to create work. As many members of the household as possible, including women and children, are involved in income- generating activities, and they often work very long hours. Most inhabit shacks that they have built themselves in slums and squatter settlements generally lacking minimal public services. Others are less fortunate. Many millions are homeless, living on the pavements of Calcutta, or Bombay (Mumbai) or other megapolises. The important role that the informal sector plays in providing income opportunities for the poor is no longer open to debate (Todaro M. 1997). Thus the burden on the informal sector to absorb more labour will continue to increase unless other solutions to the urban unemployment problem are provided. Moreover, the informal sector has demonstrated its ability to generate employment and income for the urban labour force and is already absorbing an average of 50 % of the urban labour force (Rokowski C. 1994/Todaro M., 1997).

THE IMPACT OF RURAL-URBAN MIGRATION ON CHILD SURVIVAL

Large rural-urban child mortality differentials in many developing countries suggest that rural families can improve their children's survival chances by leaving the countryside and settling in towns and cities. This study uses data from Demographic and Health Surveys in 17 countries to assess the impact of maternal rural-urban migration on the survival chances of children under age two in the late 1970s and 1980s. Results show that, before migration, children of migrant women had similar or slightly higher mortality risks than children of women who remained in the village. In the two-year period surrounding their mother's migration, their chances of dying increased sharply as a result of accompanying their mothers or being left behind, to levels well above those of rural and urban non-migrant children. Children born after migrants had settled in the urban area, however, gradually experienced much better survival chances than children of rural non-migrants, as well as lower mortality risks than migrants' children born in rural areas before migration. The study concludes that many disadvantaged urban children would probably have been much worse off had their mothers remained in the village, and that millions of children's lives may have been saved in the 1980s as a result of mothers moving to urban areas.

The limited progress of international health programs and rural development policies in improving child health and survival in many rural areas raises the question whether rural mothers or parents can improve their children's survival chances by leaving their villages and settling in towns and cities, where modern health and social services, income-earning opportunities, superior housing, stable food supplies, and modern information on child health care are generally more available. In other words, does cityward migration represent a means for rural families to experience quicker and more pronounced improvements in their children's health and life opportunities than waiting for the benefits of national economic growth or redistributive sectoral policies to 'trickle down' to the village level? If so, and in the absence of genuine attempts by governments to improve living conditions in rural areas, a case could be made that policies and measures implemented to restrict rural-urban migration discriminate against disadvantaged children and contradict the goals of child survival expressed at the 1990 World Summit for Children. Evidence that rural-urban migration enhances child survival would also bolster the arguments of those who maintain that seasonal and long-term mobility to urban areas should be allowed and in some cases facilitated as a family survival strategy or as a means to promote national economic growth (Richardson 1989; Findley 1992).

On the other hand, the conventional belief is that rapid in-migration to towns and cities of developing countries leads not only to such well-known problems as shortages of housing, jobs and social services, and to environmental degradation (UN 1993), but also to increased threats to the health of children of migrants as well as to those of the existing, resident urban population (Bogin, 1988).

Throughout the developing world, migrant women in big cities are more likely than non-migrants to settle and remain in slums and shantytowns where basic household facilities essential for good health and survival are unavailable (Brockerhoff 1993). Furthermore, the physical process of moving and resettling in low-income areas typically exposes young children to numerous hardships — new diseases, temporary residence in crowded dwellings, separation from additional care-givers, termination or decrease in the frequency or intensity of breastfeeding — that undermine their well-being. For native children, the influx of new urbanites often brings them into contact with disease agents not typically found in modern urban environments (Prothero 1977; WHO 1991), and further strains the capacity of municipal services and infrastructure to meet their basic needs. Such pressures are recognized to be most common in 'mega-cities' originally designed to accommodate fewer than five million residents, but which now encompass more than ten million inhabitants (Brown 1987; Axelbank 1988). Evidence that rural-urban migration, on balance, exacerbates child health and survival chances would provide additional justification for current policies and measures implemented by virtually all developing country governments to curb rural-urban migration in order to reduce rates of urban growth (UN 1990).The central question in this study is whether mothers improve or harm the survival chances of their children under age two by moving from rural to urban areas of developing countries, and if so, at what stage, by what magnitude, and through what mechanisms this occurs. Where possible, reference is made to the impact of in-migration on the survival chances of children already residing in urban areas, although direct evidence of such impact is unavailable.

The study uses data collected by the Demographic and Health Surveys (DHS) project in 17 countries between 1986 and 1990 to analyse and compare the maternal migration-child survival relationship in four developing regions: sub-Saharan Africa, North Africa, Latin America and Southeast Asia. Pooled regional samples are used in multivariate analyses since most country surveys recorded insufficient vital events to reliably estimate child mortality risks at various stages of the migration process. The regional perspective is intended to identify where policies to curb urban in-migration on the basis of child health concerns are most appropriate.

EFFECTS OF RURAL TO URBAN MIGRATION

As countries pursue development and industrialization drives, the need to build urban centres become certain. It is in this pursuance that leads to somewhat neglect of the no urban centres. This calls for people in the rural areas desiring to move to the urban centres. Every man wants a standard of living better than where he is now thus; the drive to move to the urban centres becomes obvious. In the developing countries, this passion becomes so great that the movement becomes rampant. The purpose of this piece is to reflect on the serious problem of excessive rural-urban migration and rising urban unemployment and underemployment in developing countries and how to restore a proper balance between rural economic and social opportunities to curb rural-urban migration.

Rural areas are isolated areas of an open country with low populationdensity. Occupation by the rural folks is mostly primary in nature. For instance, farming, lumbering, fishing, quarrying etc. are the major activities offered by the rural people. Most of the industries are being concentrated in the urban centres. It becomes relatively hard for the rural folks to make ends meet. Since the available occupations in the rural areas are mostly primary, they offer a low profit capital and the few elite are not motivated because of the primary nature, and fewer wage offered by the jobs in the rural areas, not much to be desired. The seasonality of the primary job or the agricultural activity which they are engaged in results to seasonal unemployment and the small scale business industries are been wiped off by the demand for the technological products from the urban centres thus causing structural unemployment. This leaves much people in the rural areas unemployed in most part of the year. Rural areas are characterized with chronic food insecurity widespread and livelihoods are more vulnerable. Rural folks have limited access to basic social services, safe water, roads that are accessible year round, and electricity and telephone services. Poverty is most severe among rural farmers, who are mainly traditional small-scale producers. About six in ten small-scale rural farmers are poor, and many are women. Women bear heavy workload in addition to their domestic chores, they are responsible for about 60 per cent of agricultural production. More than half the women who head households in rural areas are among the poorest 20 per cent of the population. The fate of the youth and students in rural area is not promising. They are mostly at disadvantage to their counterparts at urban areas who have easy access to quality education, internet facilities, employment opportunities, better health care and opportunities for advancement. Notwithstanding these disadvantages, the rural folks are facing, they enjoy certain advantages. For instance, living in a rural area allows residents to enjoy the natural environment instead of having to go to recreational sites. In addition, rural people do not have to struggle with the daily stresses of urban life such as being stuck in traffic, dealing with higher rates of crime, and in many cases, paying higher taxes.

These absences of stressors can have a great effect on the overall quality of life and as one researcher notes, "People living in rural and sparsely populated areas are less likely to have mental health problems than those living in urban areas and may also be less likely to relapse into depression or mental illness once they have recovered from these in more densely populated areas" (J. W. Combs 1960). The big question is, are these advantages much enough than the disadvantages to motivate those there (rural folks) to stay there? The big response to the big the question is "NO" since its disadvantages outweigh that of the advantages. These phenomena is the main push factor causing the rural folks to move to areas (urban areas) having comparative advantage than the rural areas (thus from the rural areas to the urban centers) (Samkof, 2002).

Urban area is an area with relatively high population density that contains a set of closely related activities. Urban areas are endowed with free-standing built-up area with a service core with a sufficient number and variety of shops and services, equipped with better administrative, commercial, education, entertainment and other social and civic functions. Urban areas generally have advanced systems for sanitation, utilities, land usage, housing, and transportation. Parents in the urban areas have a number of choices available for the education of their children and can often select from a long list of both public and private school districts, which leads to the potential for better education. It is also worth noting that urban areas offer residents the possibility to choose from a range of employment opportunities at any number of companies or organizations. Aside from this, urbanites have better access to choices in healthcare as well and if they suffer from diseases they have a number of specialists to choose from in their area. Development is mostly concentrated in the urban areas which intend to facilitate interaction between people and businesses, benefiting both parties in the process. There is a widely availability of modern amenities like the internet, telephone and satellite communication facilities in the urban areas. Majority of the households in urban areas are blessed with technological advancement or products which lead to effective and efficient work done since most of the jobs are done technologically instead of man power. Urban folks are at an advantage of using these technological amenities (internet facilities and computer) to broaden the scope of their knowledge compared to their counterparts at the rural area where these products (technological products) are limited, and in most cases lacking. Again there is unlimited access to financial opportunities to the urban producer. His rural colleague being poor in nature has limited access to financial institutions. The presence of credit available to the urban producers to expand their business and engage in other ventures is one of the advantages the urban folks have over the rural folks who although have the available raw materials, lack the capital and technological expertise which can be provided by the financial sectors. As stated earlier these advantages compel most of the rural folks to move to the urban centers to have access to the numerous advantages endowed in those areas.

The act of moving from the rural area to the urban centers for economic and social opportunities is term as rural-urban migration. Rural-urban migration has led to urban areas in developing countries to witness a rapid increase in urban population concentration within the past decades. In Ghana, for example, the urban population in1960 was 23 percent of the total population. The proportion increased by almost10 percent to 32 percent in 1984 (Statistical Service of Ghana, 1988). This has engendered both scholarly and policy interests and concerns. The concerns have emanated from the fact that the bulk of the urban population is constituted of migrants from rural areas and other small towns.

The loss of rural population to urban centers is often bemoaned for its implied adverse effect on rural development. The empirical data have revealed that migration is generally selective of the young, versatile, and/or better educated members of the community (Ritchey, 1976; Browning Ritchey, 1976; Browning & Feindt, 1969). This creates a form of "brain drain" on the rural populations.

· The deleterious consequences of rural-urban migration has led to the demand for urban socioeconomic amenities exceeding their supply, the urban areas often become spectacles of multifarious problems such as overcrowding, congestion, inadequate housing, high rates of unemployment and underemployment, crime and other forms of delinquency. The quality of life at both the areas of origin and destination are, therefore, significantly affected. The repercussion of the ongoing analysis is being discussed below;

Unemployment in The Rural Areas

In the developing world, there are limited job opportunities available, so the influx of people from the rural areas to the urban who are mainly unskilled in profession causes them not fit for the available jobs in the urban which are mainly skilled in nature. Thus there will be a rise in the unemployment and the underemployment rate in the urban areas as well as the rural areas. As the migration happens, the labour force in the rural areas is being reduced and this will go a long way to affect the output that can be produced in the long run.

Low Productivity in Output and Thus Low GDP

As there are more and more people leaving the rural areas which happen to be the fertile grounds of development of any developing country, there become a limited number of labour forces in the rural centres. This goes a long way to reduce the amount of produce that can be produced to feed the whole nation. The fact that most developing nations rely on agriculture and primary production for their development means that a reduction in agricultural and other primary activities means a reduction in the nation output and hence a low gross domestic product or national income. This is the reason why most developing countries with high rate of rural urban migration has huge balance of payment deficit. This is because the nation is not able to grow enough food to feed her citizens so she has to borrow from outside source to cater for her expenditure which is most at times centered on the urban communities.

Overcrowding and Pressure on Social Amenities

If a country is advanced in infrastructure and basic social amenities, there is no point to classify it as developing. To this point, it is common to see most developing countries having limited social amenities and infrastructure. This may include schools, roads, places of convenience, pipe borne water among others. As people troop from the rural to the urban, they increase the usage of these limited amenities in the urban areas. This leads to massive and quick deterioration of the amenities and hence an increase in the cost of repairs and maintenance and thus government expenditure. The increase in the population in the urban areas relative to a constant land area leads to overcrowding at a particular area and most especially the urban areas.

Increase in Social Vice and The Development of Slums

When people move from the rural to the urban areas, they have in mind that they are coming to work and feed their relatives at home. It becomes a new story when their quest for employment becomes an illusion as there is no job to absorb them. To make ends meet, most people who migrate to the rural areas resort to stealing, armed robbery, prostitution, "drug pushing" and as they cannot afford a decent house to buy, they settle in a particular place and build houses made of wood, aluminum slate among others. This community tends to be filthy all the time and not worth staying in. a room can inhabit as much as 10-15 people and it is even rumoured in Ghana that some people run a shift system type in sleeping under the same room; "sleep in the morning, i will work in the night and I will sleep in the evening and you will work at night" - what a drama. Notable areas in Ghana are Nima, Jamestown, and Chorkor, all in the suburb of Accra and Makoko in Nigeria. This point goes a long way in increasing infectious diseases and other related sickness.

Slowdown of Development in The Rural Areas

As the youth who are the agent of change in the rural areas move to the urban areas, they leave behind the aged and the too young who cannot effect any meaningful change or development in the rural areas. This leaves most rural areas in the developing countries almost the same in social setting over decades and jubilees. The aged think of what they will feed on till they die thus are not ready to cause any development, the little ones on the other hand look forward to be helped. Therefore those in whose hands lie the catalyst of change move to the urban centres in such of greener pastures. The feeble ones left at the rural areas cannot work to bring massive change which comparatively could be done by the energetic ones.

The key to solving this problem of excessive rural urban migration in restoring a proper balance between rural economic and social opportunities are;

Provision of Basic Social Amenities

The governments in the developing countries should implement policies that will enhance the provision of basic social amenities like electricity, portable water, recreational centres, schools, good roads among others in the rural areas. The government of Ghana's initiates concerning the rural electrification project is a typical example concerning this point. As these amenities are provided, they enhance healthy living and opportunity to know what is going on in the world. The recreational centres would be a source of entertainment and relaxation for the dear hardworking farmers and their relatives. To walk for miles on a third class road which neither human nor vehicle can access during a heavy rain just to attend a basic school and do the same thing when coming back is more or less a hectic and a propeller of rural urban drift. People in the rural areas should be provided with schools at least from the crèche to the senior high school level. This will limit their movement to the urban areas in search for formal education. Concerning the road, most produce have to lie idle in the farm which sometimes get rotten, therefore providing good roads will ensure fast and effective conveyance of produce from the farm to the market and thus a high value for their produce which in the long run improve their living standard.

To summarise this point, the provision of these basic social amenities will enhance active enrolment in the classroom, fast movement to the market, enhanced living condition among others.

Improvement in The Quality of Education

In an attempt to reduce congestion in urban school as well as some fortunate rural areas, there should be a policy that will lead to the improvement in the quality of education in the rural areas. This should not just be the number of school built but also the quality of teachers, the resources in the school like library, playing field, serene environment and even better structure of the school as far as building is concerned. Well trained teachers must be posted into the rural areas and adequately motivated; teachers bungalows; rural teachers allowances, teachers that will teach what they have to teach and inculcate in their student the passion to believe in themselves. The school should be of the same standard as the ones in the urban centres; a building and not just a structure. The resources available should be made ease to the student so that they can increase their innermost endowments. There have to be changes in the curriculum, teaching learning materials and the teaching strategy, a total revamping of the educational sector and changes in the attitude and the mindset of every member of the community: the learners, the educated, parents and society at large. This policy will affect the learned and later impact on the farmers and the rural community at large.

Creation of Credit and Loan Scheme

Especially, farmers and women in the rural areas should be provided with credit facilities and loan opportunities. The fact that "rural" is poor means that these loans should not bear high interest rate. Government should made available loans and credit schemes to the "rural" to expand his farm, business and buy new crops. Again subsidies should be given on fertilizers and basic farm inputs. The government can purchase farm inputs and sell it to the farmers or rural folks at a subsidised rate on credit to be paid in installment, rural banking services and rural microfinance institution in the rural areas to improve their savings and enhance their endowment and introduce them to the financial sector. The agricultural extension officers should increase their outreach programmes to educate the nation builder in the rural areas new farming methods. Women should be given credits to start or expand their business. All these should be done in a carefully studied manner so that those who really need it would be given. The main idea of this policy is to equip the rural folk to be more vibrant and productive in their field of work and prevent them from moving to the urban centres to have access to the financial institution. Therefore there is a point in granting these loans to people who will effectively use the money to improve their jobs so that when the loans are being taken, there will be a possibility of repayment.

Industrial Modernization

The government should also embark on the establishment of industries, factories and Agro businesses in the rural areas. These industries would serve as a means through which the primary produce can be processed into semi or finished goods. Example the establishment of Akomadan Tomato factory in Ghana has made it possible to process fresh tomato into canned ones; preventing loss of farm produce thus providing a ready market for the farm produce. Further, the government should build silos and other storage facilities into which farm produce would be bought directly from the farmer at a more competitive price and be stored in the storage to feed the available industries and even for export. Also, the government can also help private investors by giving them tax exemption and other incentives to motivate prospective investors to invest in the rural areas. This will go a long way to create employment to the rural folks thereby limiting their desire to move to the urban centres for employment opportunities.

Technological Sophistication

Because simple tools are used for farming, rural farmers cannot produce in large quantities, therefore providing technologically advance input like mower, ploughs, tractors, fertilizers and education in modern farming techniques among others would increase the amount of output and therefore even their wealth. This is because provision of such opportunity can positively impact on the rural folk because they can embark on large scale plantation or commercial farming to increase yield and output and their income. This will make farming to both adult and youth in the rural area attractive hence limiting their desire to get to the urban centre for other employment opportunities. Also the access to internet and telecommunication can go a long way to curb rural urban migration because the youth's access to these things would make them content in their community and thus they would see no difference in the rural and the urban centres since they can access even in the rural areas as it is in the urban areas.

***

BIBLIOGRAPHY

- Banerjee. B. (1983) "Social Network in the Migration Process: Empirical Evidence on Chain Migration in India." Journal of Developing Areas.
- Banerjee. B. (1986) Rural-Urban Migration and the Urban Labour Market: A Study of Delhi, Delhi: Himalaya Publishing House.
- Bennett KJ, Olatosi B, Probst JC. Health Disparities: A Rural-Urban Chartbook. Columbia, SC: South Carolina Rural Health Research Center;2008: 17.
- Bushy A. Rural Minority Health Resource Book. Kansas City, MO: National Rural Health Association; 2002.
- Caldwell, J.C. (1969) African Rural-Urban Migration: The Movement to Ghana's Towns, The Australian National University Press.
- Dasgupta. U. and Laishlcy. R. (1981) "Migration from Village," Economics and Political Weekly.
- Ellman, A. (1981) "Rapid Appraisal for Rural Project Preparation", *Agricultural Administration* 8, 463-471.
- Harris. J.R and Todaro. MP (1970) "Migration. Unemployment and Development A Two-Sector, Analysis." American Economic Review, 60.
- Having KM, Hale D, Lautar CJ. Ethics committees in the rural Midwest: exploring the impact of HIPAA. J Rural Health. Summer 2008;24(3):316-320.

- Joseph, Joni, "Mass Media & Rural Development" (1997)
- Kumar, Kewal.J., "Mass Communication - Critical analysis" (1994)
- Long KA, Weinert C. Rural nursing: developing the theory base. Sch Inq Nurs Pract .Summer 1989;3(2):113-127.
- Mazumdar and Mazumdar (1976) Rural Migrants in an Urban-setting. Delhi Hinndustan Publishing Corporation.
- Nelson W, Weeks WB. Rural and non-rural differences in membership of the American Society of Bioethics and Humanities. J Med Ethics. Jul 2006;32(7):411-413.
- One department serving rural America: HHS Rural Task Force report to the Secretary. Washington, DC: Health Resources and Services Administration, US Dept of Health and Human Services;2002.
- Roberts LW, Battaglia J, Epstein RS. Frontier ethics: mental health care needs and ethical dilemmas in rural communities. Psychiatr Serv. Apr 1999;50(4):497-503.
- Rural health fact sheet. Health Resources and Services Administration, US Dept of Health and Human Services. Available from: http://www.hrsa.gov/about/factsheets/orhp.htm. Accessed Jan. 19, 2009.
- Sainath, P. It's the Policy, stupid, not implementation, India Together, Sun 28 Oct 2007. http://www.counterpunch.org/sainath07212006.html - accessed on October 28, 2007
- Sexena. D.P. (1977) Rural-Urban Migration in India: Causes and Consequences, Mumbai: Popular Prakashan.
- Sivakumar, P.Sethuraman(1998): "NITs a boon for farmers" published in Communicator, April-June, 1998 Vol.32, No.2
- Srivastava, S.L. (1968) "Impact of Emigration on the Relations in a Village in Eastern U.P.," Journal of Social Research, 11.
- Todaro. MP. (1970) "Migration. Unemployment and Development: A Two Sector Analysis." American Economic Review. Population Reports. 10.1983.
- Uninsured and underinsured frequently asked questions. Rural Assistance Center. http://www.raconline.org/info_guides/insurance/uninsuredfaq.php. Accessed July 2, 2009.
- Zachariah. K.C. (1968) Migrants in Greater Mumbai, Mumbai: Asia Publishing House.

★★★

INDEX